The Science of Fitness

Power, Performance, and Endurance

I10634504

The Science of Fitness

Power, Performance, and Endurance

Greg LeMond

Mark Hom, M.D.

AMSTERDAM • BOSTON • HEIDELBERG • LONDON • NEW YORK
OXFORD • PARIS • SAN DIEGO • SAN FRANCISCO • SINGAPORE
SYDNEY • TOKYO

Academic Press is an Imprint of Elsevier

Academic Press is an imprint of Elsevier
32 Jamestown Road, London NW1 7BY, UK
525 B Street, Suite 1800, San Diego, CA 92101-4495, USA
225 Wyman Street, Waltham, MA 02451, USA
The Boulevard, Langford Lane, Kidlington, Oxford OX5 1GB, UK

British Library Cataloguing-in-Publication Data
A catalogue record for this book is available from the British Library.

Library of Congress Cataloging-in-Publication Data
A catalog record for this book is available from the Library of Congress.

ISBN: 978-0-12-801023-5

For Information on all Academic Press publications
visit our website at http://store.elsevier.com/

Working together
to grow libraries in
developing countries

www.elsevier.com • www.bookaid.org

Front cover photo attributions:
Main image: Greg LeMond racing in the Alpe d'Huez stage of the 1991 Tour de France.
Credit: Steve Selwood.
Second image: Runners competing along the James River in Richmond, Virginia.
Credit: Alan Cooper.
Third image: Karen Smith, master kettlebell instructor. Credit: Tricia Carter.

Author Biographies

Greg LeMond is a legendary and pioneering bicyclist, three-time winner of the Tour de France (1986, 1989, and 1990), three-time World Cycling Champion (1979 junior, 1983, and 1989), author of "Greg LeMond's Complete Book of Bicycling" (1988, with Kent Gordis), founder of the LeMond Fitness company, bicycle technology innovator, and fitness expert. Greg LeMond runs his company at GregLeMond.com, is a leader in the latest training equipment and power (watts) training devices, and has recently reintroduced his brand of LeMond carbon fiber bicycles. He contributes in many cycling periodicals, including Cycling News. He recently joined Eurosport as their Global Cycling Ambassador and as a TV sports commentator. Greg is uniquely qualified to describe the importance of mitochondria during the height of his racing career and also when mitochondrial myopathy resulted in his premature retirement from racing. He will explain why the modern athlete needs to know the science of fitness.

Dr. Mark Hom, M.D., is a Johns Hopkins University trained biologist (B.A. in Biology with Honors), an award-winning artist (Kenan Art Award, first place), an award-winning medical illustrator while attending medical school at the University of North Carolina (a UNC Dean's grant recipient and also a UNC Holderness Fellowship recipient), a practicing interventional radiologist (one of the most in demand medical specialties), an educator of young doctors (an Assistant Professor of Radiology), and an avid fitness cyclist (riding 4,000 miles per year). Dr. Hom will explain how the human body, various organ systems, and individual cells function in the biologic process of exercise. Dr. Hom will show that mitochondria are at the center of cell energy production, metabolism, and athletic performance, as well as an underlying cause of degeneration, aging, and many diseases. Optimizing mitochondria not only maximizes performance but also improves vitality and overall health.

Foreword by Charles A. Mohan, Jr.

CEO/Executive Director,
The United Mitochondrial Disease Foundation

Here is yet another book about exercise. Combine that with a discussion about biology, and you just might be ready to set this book down! But what if I told you that these tandem topics will provide you with the component you probably missed in all the previous exercise books you have ever read. It all boils down to one word: MITOCHONDRIA!

Let me start from the beginning. I graduated from a childhood 26" Huffy Cruiser to my first "adult" bike in 1972 when I bought a Peugeot UO-8, 10 speed. This was the same year I started teaching at a local high school. Between my new teaching position and working evenings at our local family restaurant and bar, my new bike spent more time in the garage than on the road.

Realizing the limited amount of time I spent riding, I settled on a motorized compromise. I donated my Peugeot to my brother and bought my first motorcycle. I figured at least with a motorcycle I could combine riding with necessary transportation back and forth to work.

Fast-forward to 1995. Within a year that Greg LeMond announced his retirement from bicycle racing due to mitochondrial myopathy, I lost my daughter to a mitochondrial disease. She was only fifteen years old.

The loss of a child can immobilize distraught parents for many years, or it can fuel in them a passion, a cause, a quest, to somehow make sense out of indescribable tragedy. In my case, for the past twenty years, I have focused my energies on mitochondrial dysfunction and disease – what it is, how it is diagnosed and treated, and ultimately, to help find a cure.

MITOCHONDRIA. These same cellular organelles that due to a mutation malfunctioned in my daughter are also the ones that fuel your ability to breathe, to eat, to exercise, and to live. Mitochondria are responsible for producing 95% of the energy your body uses every day. They complete the final steps in extracting that energy from the last meal you ate. Dysfunctional and abnormal mitochondria can result in life-threatening diseases, and are a key component in the aging process itself.

And that's where *The Science of Fitness* enters the picture. Greg LeMond and Dr. Mark Hom demonstrate that fitness and mitochondrial health are inextricably linked; you cannot have one without the other. Exercise and proper nutrition can increase both the function and number of mitochondria, resulting in more energy and stamina.

If you want to have an edge on your athletic competition, or just desire a healthier and longer life, read *The Science of Fitness*!

Charles A. Mohan, Jr.,
CEO/Executive Director,
The United Mitochondrial Disease Foundation (UMDF)

The UMDF Mission is to promote research and education for the diagnosis, treatment, and cure of mitochondrial disorders and to provide support to affected individuals and families. For more information on mitochondrial disease visit www.umdf.org

Preface by Greg LeMond

I knew I was dying. Blood poured from my wounds, spurting wildly from my neck. I was trying to cry out, but with every breath I choked and gargled through the blood in my throat. Inside, I repeated over and over: *Oh, my God, I'm shot!* My strength seeped away into the scrub and brush of the remote northern California ranch.

The shotgun blast was unbelievably loud, right in my ear. My brother-in-law accidentally fired just as I stepped out from the cover of a huge wild raspberry bush in front of him; then all I knew was noise and confusion. I thought my own gun had gone off, and I stared at it, utterly confounded. Our small family hunting party was forty miles from the nearest hospital, and I was bleeding to death. Oddly, I was not feeling any pain, but I wanted so badly to tell the others to get me an ambulance; I needed help, I was dying here.

It seemed hours until the California Highway Patrol helicopter arrived, although I learned it was more like 20 minutes. Paramedics secured me to the outside of the two-man helicopter. I watched the rotor blades cutting across the clear April sky. What a strange, strange way to go, gazing into the heavens.

We landed with a thump at the University of California Davis Medical Center in Sacramento. The last thing I remember was the emergency room physician slicing into me and inserting a chest tube. Pain had made itself known. I was not conscious again until some 6 hours later, after a team of diligent surgeons saved my life.

I lost sixty-five percent of my blood volume in the field. Shotgun pellets had ripped through my back and side, puncturing my lungs, liver, kidneys, and the lining of my heart. Despite the surgeons' best efforts, more than three dozen lead pellets remained embedded in my upper torso, including three in my liver and two in my pericardial lining. But I was alive.

My first athletic love was not cycling, as some believe, but freestyle skiing. I grew up in Reno, Nevada, a bit of a wild child who spent as much time as possible outdoors. Tearing down the ski slopes of the Sierra Nevada Mountains was exhilarating. I found that I loved pushing myself, testing my limits. In 1975, I attended a freestyle ski camp run by my hero Wayne Wong, who suggested I try cycling to stay in shape and build muscle onto my 14-year-old frame during the off-season. Bicycling is popular with skiers because it strengthens the same hip extensors, quadriceps, and calf muscles required in competitive skiing. I was sold on cycling as a way to exercise, and soon found it had many benefits for me beyond the physical.

I was a boy who just could not sit still. I had trouble focusing in school. Parents and educators then did not have the skill set to diagnose and cope with what we know now was a classic case of Attention Deficit Hyperactivity Disorder (ADHD). ADHD certainly was not the frequently medicated childhood disease it is today. My triumph over the symptoms was found atop two thin tires over many dusty miles. Even an hour of exercise cleared my head and sharpened my focus. I was transformed.

I might never have become a racer if the Sierras had enjoyed a typical snowfall that winter. Instead, there was a drought. Snow was scarce. A guy at the local bike shop asked if I had ever thought of entering any bike races. Why not?

Soon I was winning the races I entered for fun, even beating Olympic and World class racers. At the age of 15 years, with seven months of cycling under my belt, I felt confident in my abilities, was told I had an impressive natural talent, and felt I had found my place. And I want you to know that I did not come from an athletic, competitive family, unless you count bowling and trap shooting among your World class sports. My sister, Karen, was an incredibly talented gymnast, eventually ranked first in the nation, but we were certainly the exception to the rule among the LeMond's. Some observers credited my immediate success with luck: the *luck* of growing up at high altitude, developing powerful lungs; the *luck* of youth; the *luck* of stumbling into a sport I was good at. All of that may be true, but I know now that I was blessed with genetic gifts, an insatiably curious mind, a talent for strategic thinking, and the guidance of some of the most amazing coaches in the world.

When I was just 17 years old, I made a list of cycling goals for myself. They were certainly lofty goals, by anyone's standard, but even as a teenager I knew I possessed the dedication and skill to meet them. Here is my list from 1978:

1. Win the Junior World Cycling Championship
2. Bring home the Olympic gold medal in 1980
3. Win the Professional World Cycling Championship by 23
4. Win the Tour de France by 25

I met each of these goals except one, winning an Olympic medal. The United States boycotted the 1980 Olympic Summer Games to protest the Soviet Union invasion of Afghanistan, and global politics thwarted my Olympic dream. I do not dwell on the prize I did not win, the race I lost, or any of the personal setbacks I have suffered. That is just not who I am. When I think about the clarity and confidence I had in myself at the age of 17 years, I just have to smile. I believed in myself, believed in the counsel of expert advisors, and, of course, in my own *luck*.

A year later, I advanced to the world stage of bike racing. I met my first listed goal at the age of 18 years, taking the gold, silver, and bronze medals at the 1979 Junior World Championships in Argentina. I was a dogged researcher into the sport I loved. I wanted to know everything, understand all aspects of competitive cycling. I read everything I could get my hands on about my sport, this slightly strange, foreign obsession with speed and endurance on a bicycle. Few Americans pursued this sport, and I had not heard of the Tour de France until I was 16 years old. I wanted to know how and why my body could perform like this, and I created my own unofficial field of study in sports physiology. I know that I was among the most knowledgeable riders, I took nothing for granted, and I gleaned information everywhere I went.

Even so, it was many, many years before I knew why I could race a bike so well.

Six months after winning the Junior World Championships, I was competing in France with the US National team in preparation for the 1980 Olympics, just 4 months away. Our team entered one of the two "open" races in cycling, that is, open to both professionals and amateurs. I won that race and was quickly approached by some of the best teams in the world. With several offers to turn professional in Europe, I carefully weighed my options. I chose the French Renault team for its coach, whom I (and many others) considered the best in the world, Cyrille Guimard. This was the man I wanted to take me from amateur championship to World class competition. (Not only did Renault have Guimard, it was led by three-time Tour de France winner Bernard Hinault.)

Cyrille Guimard was the best coach I ever had. He taught me about cycling and the human body. In fact, he was the first professional cycling coach who attended school to study the physiology behind the training process. His racers would not train according to any arbitrary, time honored systems; we trained with purpose, including plentiful rest periods. He was fascinated by aerodynamics, and took us to our sponsor's Formula One auto racing wind tunnel in Paris. The equipment, positioning, and training protocols that our Renault team developed 30 years ago are still considered revolutionary today. One of the 2014 Tour's top professionals told me that he still follows my training methods, described in a book I wrote in 1985.

Coach Guimard watched over my career, implementing a long-term plan to develop me into, as he would say, a legendary rider. Upon becoming a professional, I underwent many tests to see how my body dynamics compared with other athletes. VO_2 max, or maximal oxygen consumption, is the most accurate laboratory test of aerobic fitness. It measures how much oxygen you can transport to your cells with your cardiac output as well as how much oxygen your muscles can consume. VO_2 max testing is crucial to anyone in serious training, and the gold standard to determine an athlete's potential.

My VO_2 max result was 93, one of the highest ever recorded. By contrast, an average male typically tests in the 40s. An average athlete's score may range in the 60s, exceptional world class cyclists test somewhere in the mid-80s. My number meant that I had an extremely high potential for athletic performance. Improving that tolerance improves performance. It was clear that I had an obvious physical advantage other over athletes, an advantage that is part of my genetic makeup.

Anyone can improve their VO_2 max with proper training. Increasing your oxygen usage will increase your stamina. Think of the human body as a furnace: fuel goes in, oxygen goes in, and energy comes out. Oxygen combines with food, turning it into active energy. The better your fueling and oxygenation, the more efficient your furnace. During my racing career, I took very good care of my furnace by eating right, sleeping right, and training smart.

Cyrille Guimard knew what he was doing and together we achieved great results. In 1983, I happily crossed the third goal from my list, winning the World Championships. Then, in 1984, I finished the Tour de France in third place, despite suffering from stubborn bronchitis. The next year, I finished just behind my former teammate Bernard Hinault in his fifth and final Tour victory.

In 1986, I won the Tour de France. I was the first and still the only American to claim the Tour de France trophy. The thrill was absolutely incredible, far greater than I could have imagined 7 years earlier when I inscribed my personal goals. I had met them all—all that I possibly could—and at the age of 25 years, I was on top of the world.

Nine months later, I was clinging to life in Sacramento, California.

My recovery did not go smoothly. The trauma of the shooting affected me in myriad ways, and I was not alone in my suffering. The hospital staff called my wife Kathy, informing her that I had been shot. I do not know what kind of odds they were giving me, but the news propelled Kathy into early labor with our second son, Scott. Not long after I hit the recovery room, Kathy's obstetrician fetched her from UC Davis Medical and took her to Sutter Memorial Hospital, which specialized in high-risk deliveries. She was only 8 months pregnant, and had been in labor all day. A very hard-won 19 days later, Scott was safely delivered in perfect health. His weak father was wheeled into the delivery room just in time to see him cross the finish line.

At the time I was shot, I weighed 150 pounds, and had a body-fat index of just 5%. I was extremely fit, but the trauma and the catabolic effect that kicked in when I was survival mode caused a loss of nearly 30 pounds of muscle mass. I was terribly weak.

Before the end of summer, I needed more surgery, this time to repair intestinal blockage from internal scar adhesions. That is major surgery any time, much less so soon after my life-saving operation. Of course I endured days of despair, but I was so grateful to be alive that I concentrated on that good fortune.

There I was, the acknowledged best bicycle racer in the world, rapidly losing the physical accomplishments of my months and years of grueling training. Competing in the Tour de France has been likened to running a full marathon every single day for 3 weeks. The sheer brutality of the experience is hard to fully convey. Now, at the age of 25 years, at the pinnacle of my success, I was stretched out in a hospital room, wondering if I would ever be able to get on a bike again.

My French team sent me a letter expressing their relief that I had survived the shooting, and not-so-gently letting me know that my services would no longer be required. Kathy and I had been married for 6 years, had two sons, Geoffrey and days' old Scott, and had gritted our teeth through some tough early years in France. I struggled to concentrate on getting well, not on the difficulties facing our little family. It was a long time before I could walk even a 100 ft without stopping to rest.

As I got stronger, my determination came back. I was going to reclaim my rightful place in the cycling world. It was going to take every single ounce of stubbornness and grit that I possessed, and every single thing I knew about body mechanics and proper training to find my way back to the Tour. The road back to Paris would be the greatest challenge of my life.

Kathy and I decided that my best chances lay with a new team contract for the 1988 season, and that no one needed to know the full extent of my injuries. As soon as I could pass a team's required physical fitness tests, and had the rather uninformed approval of my surgeons, I could get back to work. By August, I had a contract…with stipulations. I had to enter a professional race by the end of the calendar year.

I realize now that I returned to training much too quickly and much too hard. I was pushing an envelope that no one had so much as recognized, to say nothing of besting my very real limits. I entered a 60-mile race. After just 1 mile, I had nothing left. I might as well have had two flat tires and a broken frame.

Now I pulled together all that I had learned since entering my first race at the age of 14 years. I had to strip my training to its most basic elements. I could not ride as far or nearly as fast as I had then. I simply had no fitness. In spite of my gifts, my natural talent, and my mighty VO_2 max, I had to wonder if my luck had run out. I was being dropped off the back of races by riders I had routinely beaten. I could not offer any excuses either because to blame my injuries would have jeopardized my employment, now and in the future. I had to do my best with what I had. I had to shut up and ride. No athlete had ever come back from the type of injury I had sustained, and it looked as if I would not be the first.

This frustration and futility went on for 2 years. I could not make myself perform the way I wanted to, the way I knew I could. I had always been in the front of the peloton, and now I was in the back of the pack. I knew I was being laughed at. Riders avoided making eye contact with me. It looked as if I was finished, and I was embarrassed and alone.

I entered the 1989 Giro d'Italia under a crushing load of personal stress. The race began like every other the past 2 years; I was in the back of the peloton. This precursor to the Tour de France, spanning some 22 days, was taking on the newly familiar texture of failure. One night I called Kathy and told her I just wanted to come home, to be with her and the kids, and to quit racing. She asked me if I had given it everything I had, so that if I were to retire, I could do so without regrets. If my answer was no, that I had not given my all, then she wanted me to stick it out until the end of the season. Either way, she promised to fly over as soon as she was cleared by her doctor.

That conversation was pivotal. I felt the pressure lift away from me. I would be all right, even if I did not win. If I just focused on doing my best, that was enough to keep me going.

My mood was so improved. I felt stronger and stronger, and by the last 33 mile time trial stage, I pushed myself as hard as I could. To everyone's surprise, I finished that stage in second place, ahead of Laurent Fignon, my former teammate and current greatest rival. One month before the Tour de France, I felt like my old self again. I was not thinking about winning; I would have been thrilled with a top-ten place. After my shaky recent performances, I was not sure what my body and spirit could accomplish. Cycling analysts looked at my Giro d'Italia results and figured that I did not stand a chance.

During the 1989 Tour de France, Fignon and I battled through the mountains. I was in good form. The Tour was set to end in an unusual time trial finish in the heart of Paris.

I believed I could win the time trial, but could I win by enough time to actually win the Tour? All I knew for sure was that it would be close, incredibly close. My decision now was to tackle that time trial with everything I had in me, maybe with everything I had *ever* had in me. I asked not to hear my split times. I did not want anything to distract me.

Mile for mile, that short time trial was the ride of my life. I beat Fignon's time by 58 seconds, enough to win. In what many believe was the closest, most dramatic Tour finish in history, I won the Tour by a mere 8 seconds. My joy in this moment was so complete because I had finally defeated my worst enemy: my own doubts and fears. Clichés collided in my head, ringing with truths about never truly tasting victory without having experienced bitter failure. Everything I ever knew about myself, about cycling, and about competition had come together. The view from the top of the world is spectacular, especially when you have been so low that you have been blind.

I went on to win the 1989 World Championships a few months later capping my remarkable comeback season. Then, incredibly, I won the Tour de France again in 1990. That was truly one of my sweetest victories. I was surrounded by supportive teammates and a team owner that I will never, ever forget. We had found a perfect blend of our personalities and talents, a deeply satisfying and effective way to win the toughest cycling event in the world. There are times I wish I could have gone on to win more Tours and other big races for the wonderful Z Team.

Miracles do happen, especially if you brace them in determination and all-out effort. But few miracles last indefinitely, and the last few years of my career were plagued by an impediment I could not have foreseen. I was 30 years old by 1991. This should have been the prime of my racing career, but my performance times began dropping away. It was clear to me that my body was not operating at its peak. I was exhausted, and had difficulty recovering after races. Once again, riders I had easily defeated were coasting past me. The old doubts came back: Was I training too hard? Not hard enough? What could explain this decline? My VO_2 max tests now showed that I was taking in much less oxygen after training and racing. How could I have come back so strongly, only to face this disheartening fall?

Finally, I turned to a specialist in Minneapolis. He took a muscle biopsy that revealed ragged red fibers inside my cells. This was mitochondrial myopathy, he said. No doubt about it. I had not heard of this disease, and did not know anyone who had.

As I now know, I have an extremely rare type of mitochondrial disease. Cases are either genetic and therefore permanent, or temporary caused by an outside stimulus. Mitochondria are organelles found inside each of our cells, where energy is produced from fuel and oxygen. My mitochondria were being poisoned.

The shotgun pellets lodged in my body were slowly leaching lead into the very dynamos of my cells. The doctor explained that mitochondrial myopathy leads to muscle weakness, breathing difficulties, and an inability to endure long stretches of physical activity. I had the misfortune of being exposed to a poisonous toxic stimulus that was now all throughout my system, and permanently damaging my energy metabolism. No wonder this was affecting my ability to race! The devastating effects of this poison might not have become so noticeable had my decline not coincided with a new and dirty little secret in cycling. Many racers were now involved in the secret, illegal use of a performance enhancing drug called synthetic erythropoietin, or EPO. The drug boosted their performance to an extent previously unseen. Their benefit added to the mystery (and misery) of my decreasing ability to compete. The playing field was not even; it had, in effect, been pulled out from under me. I would not and did not, under any circumstances, ever ingest or inject any illegal performance-enhancing substance.

In 1994, I retired from professional cycling. I was 33 years old.

Twenty years have passed. I am a very lucky man, after all. Unlike some mitochondrial disease sufferers, I am not in a wheelchair, and I am very much alive and kicking. I can do all of the things most men and women my age can do. I have learned, though, that more than an hour of exercise causes the lead levels in my blood to rise significantly. I still ride my bike, play tennis, enjoy fly fishing, and generally keep up with regular physical activities. I will not be pedaling up the Alps any time soon, but as my kids say, I have been there, done that, and worn the yellow jersey.

Two years ago I received an email from a doctor in Virginia that grabbed and held my full attention. Dr. Mark Hom told me he was writing a different kind of fitness book, one which, he said, was inspired by the story of my career. He could explain physical performance through the function of mitochondria. He could answer so many of the nagging questions I had about my cycling successes and failures. Naturally, I was intrigued. I sat down and wrote a 4,000 word email response, about as long as this preface. All the thoughts and emotions I had felt over the years came rushing back to me, and I felt compelled to share them with other athletes by coauthoring this unique book. Mainstream fitness publishing was saturated with trendy, superficial fitness books that barely scratched the surface of the science behind performance. The science behind my initial meteoric rise, devastating injury, inspiring comeback story, and puzzling decline will definitely help anyone, veteran trainer or novice athlete, understand mitochondrial biology, the most significant factor in exercise, fitness, diet, and health.

Although scientists have long been aware that mitochondria make our energy, the true extent of their importance in fighting disease and improving general health is just being realized. So many prevalent diseases of the modern age, including obesity, type 2 diabetes, and metabolic syndrome, can be prevented and dramatically improved with exercise. There is no pill, no drug, no medicine available that can do for you what 1 hour of exercise can. Dr. Hom and I are excited to share a world of potential with you. We explain just about everything a modern athlete needs to know about getting fit naturally and scientifically. I think the practical ideas we present can be a huge help in making yourself the best athlete you can be.

I want you to unlock your full potential, just as I was able to do. You do not have to be the fastest, the strongest, the toughest, or even the smartest athlete in the world. But maybe you can be, with the right knowledge, enough willpower, and a little luck.

Greg LeMond

Preface by Dr. Mark Hom, M.D.

One of the earliest and fondest memories of my childhood was learning to ride a bicycle. The freedom, mobility, and expansion of my world were new and exciting experiences at that highly receptive and impressionable age. As a young boy growing up in Maryland, I was fortunate to have forests, creeks, lakes, and bike paths to explore. Montgomery County where I lived was a suburb of Washington, DC and the home to many government employees whose taxes supported some of the best public schools in America. My schools had innovative curricula that focused on exposing students to mathematics and science at a very early age and also emphasized daily exercise.

My first bike was a blue Raleigh "Chipper" a single speed bike with a rear coaster brake, banana seat supported by a sissy bar, a kicked out front fork, and "chopper" style chrome handlebars. Designed to mimic the Harley Davidson motorcycles from the movie (1969), "Easy Rider," Columbia Pictures, that first bike vastly extended my world, well beyond my neighborhood. The training wheels were quickly discarded because to turn on a bicycle you have to lean and counter steer. With a tricycle or training-wheel bike you steer in the direction of the turn, stay upright, and concentrate on pedaling. A better way to learn how to ride is with a pedal-less scoot bike with the child sitting on a low seat that allows flat foot contact with the ground. Forward movement and stopping are all with foot-ground contact as the child learns balance, momentum, and counter steering, the first steps in mastering a bike. The pedaling can come later. They say once you learn how to ride a bike, you never forget. That is because the brain (and the cerebellum in particular) makes new connections when you learn to ride. This rewiring of your brain or "muscle memory" can last your entire life. It was a simpler time then and kids were expected to be physically active and playing outdoors, not surfing the Internet or gossiping on cell phones. It was about this same age when I had my first exposure to cell biology in the 3rd grade. New microscopes had just been delivered to the school and I clearly remember unpacking them and using them for the very first time. The light source was a concave mirror that reflected sunlight streaming through a large window (I remember that we had to plan our microscope work around the noon hour to get the best light). That first day we used our microscopes to look at onion skin. They say odor induces the recording and recalling of long-term memory, and I clearly remember the strong smell of onions even decades later. Being one cell layer thick, the onion skin was a good demonstration of what constitutes a living plant cell (cell wall, cytoplasm, and nucleus). This was an epiphany. Before using that microscope, things were much simpler and easier to explain. I had assumed the onion skin would be featureless like a sheet of cellophane. I had already taken my toys apart and was disappointed to find them hollow and empty inside. However, this time seeing the array of cells and then the tiny parts inside each cell made me realize that the natural world was much more complex, fascinating, and worthy of study. This was probably the singular event that directed my life to the study of biology and eventually medicine.

Two years later, I wrote my first school science essay on the topic of how plants were the source of all of our food, plants recycled our water and oxygen, and plants made life possible for animals and humans. To my great disappointment, I only got a B+ because the teacher noted in red ink that actually the sun is the source of energy for plants and that I should have taken the concept this one step further. As a young child, I made a promise to myself that I would always look deeper into subjects and strive to seek out the real underlying causes. This not only applied to science but my other school subjects as well. We had a history teacher who would always make us think beyond mere rote memorization of names and dates. The very first history lesson was for us to read from three different sources about how Christopher Columbus secured the financing for his journeys from Queen Isabella. This confused us because unlike any lesson before, the material was told from three different perspectives and contained contradictory information. The teacher asked us what we thought was the point of this lesson and it was clear to me that it was more than just facts about Columbus. I raised my hand and said that the lesson was about the difference between primary and secondary sources of information, which was correct. By always trying to learn the larger underlying principles, the smaller details (that we were tested on) would fall into place and begin to make sense.

When I got to middle school, the boys' Physical Education (PE) class was divided into a high group and low group (which sounds awfully cruel by today's egalitarian standards). As a December child, I was nearly a year younger than some of my classmates, I was that much behind in adolescent growth and coordination, and I did not respond well to the pressure of team sports, so I was content when I was placed in the low group that first year. But at some point, I decided that I did not want to be a scrawny kid anymore. In shop class, we were encouraged to design our own projects, so I built a steel pull-up bar with quick-release metal brackets that I installed in the doorway of my bedroom. I trained by doing pull-ups and push-ups and by the time the yearly pull-up/push-up test was recorded in PE class, I scored 12 pull-ups and 25 push-ups, just as many as the boys in the high group. When the PE instructor called me into his office one day, I thought I was in trouble for fighting. He said that he noticed two things: that I had all A's on my report card and that I scored high on the pull-up test. That day he moved me up to the high group. Although I still could not score a lay-up with a basketball to save my life, I was the only boy I knew to move out of the low group to the high group. With my success in strengthening my upper body, I began to bicycle as more than just a way to explore but also to improve my fitness. On a bulletin board (mind you a real bulletin board, not Craigslist or eBay), I saw an advertisement for a lightly used Schwinn Paramount ten-speed bike for $350. Although it drained every last penny of my life savings at the time, I had to have it. The Paramount was the ultimate ride of its day with hand-fitted chromed Nervex lugs, Reynolds 531 double-butted steel tubing, and candy apple red paint. My best friend asked me why I was wasting my money on a bike when I would be getting my learner's permit to drive a car in a year. Looking back, it is about this age that many discard the bicycle as a childhood toy and move on to other more adult interests. They say you never forget how to ride, but unfortunately too many people forgot how much fun it can be. Undeterred, I rode every day after school and tried to get in as much training as possible before sunset. That meant strenuous, fast rides in a limited period of time. I sought out every challenging hill and climbed them as fast as I could. I remember riding so hard that my water bottles were empty, my shirt was drenched with sweat, and my legs felt like rubber bands when I carried the bike back down to storage in the basement. One day while riding I looked down at my leg and saw a swelling going down the outside of my calf. Oh no, what had I done? I looked at the other calf and it was swollen too! I later found out that I was developing my soleus muscles, the deep calf muscles normally hidden by the more obvious and superficial gastrocnemius muscles. In trained cyclists the soleus muscle in each calf becomes enlarged (hypertrophied) to be seen from the side. My quadriceps muscles had also hypertrophied and I was starting to take the shape of a cyclist. My heart was getting stronger too. As part of the warm up for soccer in PE class we would run a lap of the field before playing. Before cycling I would be towards the back of the group gasping for air and wanting the warm up to be over. After training with cycling I was now leading the other boys in the warm up and beating them to the ball in play. I continued bicycling because it always felt so natural as a way to explore my world and get stronger at the same time.

I took every advanced biology class in high school and then majored in biology at the Johns Hopkins University. I learned that the mitochondria in each of our cells generate the energy that drives muscular power and other vital functions in the body. At that level, we were taught the normal functions of mitochondria, and we wrongly assumed that they performed their task perfectly and unendingly, needing no special attention. One book stood out at the time, Lewis Thomas (1974), "The Lives of a Cell: Notes of Biology Watcher," Bantam Books, New York. With a wide view of the biosphere and specific microscopic detail, Thomas showed that all life on Earth is intimately connected, no more so than how mitochondria came to be inside of our cells. Although Lewis Thomas did not apply this synergy to sports, the energy that mitochondria supply to our cells powers our every athletic endeavor. During medical school at the University of North Carolina, we spent the first year learning about how the human body works including: the major enzyme systems, organs systems, anatomic systems, and physiologic systems (including mitochondria). The second year was devoted to disease, infections, and pathology (i.e., what happens when things go wrong). The third and fourth years were spent rotating through the different fields of medicine and surgery to put our knowledge to practical use. At that time there was a disconnect as we knew how pervasive mitochondria were in all of our major organs and we had inklings of mitochondrial genetic disorders, but compared to the more classic diseases, little was known about how to optimize mitochondrial function or how confidently diagnose and treat mitochondrial diseases.

In the period between 1890 and 1920, American bicycle racing was in its heyday with velodromes all over the country and stars such as Marshal "Major" Taylor. With the advent of the automobile (Model T Ford, 1927) bicycle transportation and bicycle racing took a steep dive. When I took up the cycling, bike racing was an obscure sport contested in faraway Europe with no local youth teams for me to join. There was not yet a modern American cycling hero of my own era. But then word got out that a blond-haired, blue-eyed American (with a French sounding name, "LeMond") was beating the Europeans at their own game. He even won the most grueling physical contest in all of sport, the Tour de France (1986). Back then we regarded sports champions as having natural born talent and/or secret training techniques that they were reluctant to reveal. But after this win Greg wrote, with Kent Gordis (1988), "Greg LeMond's Complete Book of Bicycling," Perigee Books, New York, that revealed that he was a thinking ath-lete who was willing to openly share the secrets of his success with fellow cyclists and rivals alike. Besides a big dose of natural ability, he showed that his success was also due to his applying a scientific approach to bike set up (such as proper seat height for better leg extension), pedal stroke (what really happens on the down stroke and how to keep applying power at the bottom of the stroke), how to read your body, and how to eat for performance. By virtue of having the best coaches, constantly innovating, and formulat-ing his own unique training philosophy, Greg was one of the first cyclists to employ and elucidate a systematic approach to bike setup and physical training. Through his book and by continuing to win epic races, Greg inspired an entire generation of cyclists.

Greg LeMond and I are about the same age. While I was training to be a doctor, Greg was training to be the best athlete in the world. Shortly after he became the first American to win the Tour de France, he nearly died in a shotgun accident. Like many of his fans, I thought his career was over. Amazingly, he came back and won the Tour again, in the most dramatic finish of all time (1989). He was named Sports Illustrated magazine's Sportsman of the Year, as well as becoming my personal sports hero and the inspiration for this book. Because he was considered one of the most physically fit athletes of his era (undoubtedly having superior mitochondria), I and many others were shocked when LeMond announced that his career was cut short by a mysteri-ous illness called "mitochondrial myopathy." The shotgun pellets imbedded near vital organs were leaching elemental lead and poisoning his mitochondria. With his diag-nosis, the general public (and many clinicians) began to realize that things could go wrong with mitochondria.

"We are at a crucial juncture in the perception of these disorders. The recent announcement by the American cyclist Greg LeMond that he is retiring from competitive cycling because of a mitochondrial myopathy has brought these mysterious disorders to the attention of the public.... This announcement has been a seminal event in the public awareness of the mitochondrial disorders."

Johns, Donald R. (1995). Mitochondrial DNA and disease (and correspondence). *New Engl J Med 333, 638–644.*

We now know that mitochondria are extremely sensitive to toxins, certain medications, free radicals, genetic mutation, and inactivity. Research has revealed that mitochondria respond to exercise by multiplying. In fact, mitochondria are a major explanation of fitness improvement. Due to vigorous training, an elite athlete will have many more mitochondria per cell than a recreational athlete. VO_2 max is the best lab test of fitness because it measures cardiac output (energized by mitochondria) and skeletal muscle power (also energized by mitochondria) by means of oxygen consumption. Nearly all of our oxygen is consumed by mitochondria, so VO_2 max is really a measure of the athlete's mitochondrial "engine." At the height of his racing career, LeMond had one of the highest VO_2 max's ever recorded, due to his genetic gift and superior training methods. So as it turns out, there is a lot we can do to support, protect, and maximize our mitochondria. Just as a successful shepherd tends to the needs of his flock, the better you take care of your mitochondria, the better you will thrive.

I first started writing down my thoughts about the science of fitness during a midlife crisis. My wife and I had always been healthy, but we had become complacent and less active. This was a key turning point in our lives and future health. Would we slip into a sedentary lifestyle or fight the inertia and get moving again? Instead of resigning ourselves to mediocrity and saying good bye to the vitality of our youth, we decided to become as physically fit as possible. I had always loved bicycling and tried to share this favorite sport of mine with my wife, but I made the mistake of trying to adapt one of my old bikes to fit her. What seemed like a minor point to me (her needing a bike she could call her own), made a tremendous difference in her psyche and her motivation to begin exercising in earnest. Against my advice of purchasing a female specific racing road bike, she selected a comfort cruiser bike. One ergonomic advantage it had was that she could stop flat footed while still on the seat. We found a business park with a 6 mile loop so we could ride at our own individual paces but not lose each other. I decided to start seriously ramping up my own fitness as well and started training harder and more frequently. Although I was getting fitter and faster, it seemed to be a lot more work than when I rode in my youth. I bought every fitness book at the book store, seeking that one book that would explain it all; however, that search was in vain. Each book advocated interval training, but none thoroughly explained why. As an inquisitive person and a medical doctor who was always looking for underlying principles, I wanted and needed to know why.

One weekend I led my wife on a ride downtown and we ended up at her office. I let it dawn on her that bike commuting was possible (something I had been doing for two years). In a few months of riding workdays and weekends, my wife was extending her rides to 30 miles or more and riding much faster. She was literally riding the fenders off her cruiser bike (the fenders would vibrate loose and fell off twice). At this point, we agreed to buy her a proper carbon fiber road bike, and after she learned to use cleated pedals at the office park, we joined the Richmond Area Bicycle Association (RABA.org) which is a 1000+ member bike club. With rides classified as A+, A, B, C, or D nearly every day of the week, there was further motivation to keep riding and improving. Companionship and light competition are great motivators when sticking with an exercise program. As a doctor I was able to closely observe during side-by-side riding how cyclists of different ages, sex, weight, and ability trained and how successful that training was over the seasons. Pace riders would ride at the same effort throughout a long ride and would show slow gradual improvement, but it would take many miles and many continuous hours of riding to do so and they never seemed to break through to the next tier of speed and performance. The A and A+ riders on the other hand would always push the pace, attack hills, bridge across gaps, and sprint to the store stops. As the season progressed, these riders would see major steps in performance improvement. The fast riders would get faster. An encouraging observation was that if a middle-aged rider (such as myself) trained properly, it was possible to keep up with or even pass riders half your age. Upon doing some literature research, my investigation into fitness would always come back to mitochondria, the intracellular organelles that generate our cell energy. Mitochondria constitute a large part of the fitness building process when athletes train strenuously. I decided to focus my training by doing all I could to build up my mitochondria and avoid all the things that might impair my mitochondria. It worked wonderfully as it made me to focus on the types of training that would benefit my fitness the most, created new challenges to prevent fitness plateau, motivated me to train with intensity and variety, and forced me to take better care of my body by eating a more natural (and less artificial) diet.

As a practicing physician for 25 years, I have come to realize that a great deal of disease and suffering can be prevented with exercise and a healthy diet. As a radiologist, I can see deep inside the human body using state of the art medical imaging such as computed tomography (CT scans), magnetic resonance imaging (MRI scans), positron emission tomography (PET scanning), ultrasound, and real-time fluoroscopy (X-rays). Unlike most other medical specialists, I can see beyond the skin surface and see my patients' organs in great detail. I can see the heart beating, blood flowing, lungs billowing, muscles contracting, and joints flexing. The book will begin with an organ system by organ system explanation of fitness, instead of the usual focus only on muscles. This more complete approach, in combination with our cellular and intracellular explanation of fitness, will be the most complete health and exercise book to date.

As a fitness cyclist who rides 5000 miles per year, I have vastly different exercise goals than an elite competitive athlete. I began to investigate which exercise programs would best benefit the health of my patients as well as my own health. What I discovered is not mentioned in ordinary fitness books and I hope my fellow athletes will find the information contained here just as eye-opening as I did. I am very fortunate to be joined by my coauthor, Greg LeMond. Greg draws from his experiences as a legendary athlete who won at the very highest levels of competitive sport, all the while advancing bicycle technology and creating a systematic approach to training. Although our life experiences are vastly different, we all came to the same inescapable conclusion that learning the science of fitness is the key to realizing our human potential.

Our book will explain how to scientifically improve fitness, boost cell energy, burn fat, and maximize athletic performance. It is written in such a way that athletes of all abilities can learn and benefit from our combined knowledge and advice. Just as with the onion skin experiment, fitness is more wonderful when thoroughly investigated. This is the pathway to becoming scientifically and physically fit.

Dr. Mark Hom, M.D.

Acknowledgments

There are almost as many opinions and beliefs about exercise as there are athletes, so when writing my parts of this book, I sought and received input from many people: cyclists, runners, fitness trainers, strength coaches, researchers, and experts.

First and foremost, I would like to thank my coauthor Greg LeMond who has always been my sports hero, not so much for winning but for overcoming tremendous obstacles on his way to becoming a champion. My old copy of his book "Greg LeMond's Complete Book of Bicycling" with Kent Gordis is dog-eared from decades of rereading it because it showed how much thought Greg put into his sport. Younger readers may not be familiar with his tribulations, but perhaps they will rediscover the same source of inspiration that I did. Greg's wife Kathy LeMond was very helpful in keeping us all connected during this past year when Greg was one of the busiest people on the planet. She was there during Greg's darkest days and during his greatest triumphs.

As a first time book author, I am particularly grateful for the guidance and support given by our editorial team at Elsevier. Stacy Masucci (Senior Acquisitions Editor) recognized the relevance of our scientific approach when other publishers assumed that athletes were not quite ready for a science-based fitness book. Stacy guided me through two rounds of peer review, helped shape the structure of the book, and had the insight to know that we really needed to write two books: this affordable version for athletes, students, and the general public; and also a scientific reference version due later "Mitochondrial Fitness: The Science of Athletic Energy" for scientists, clinicians, and researchers. Shannon M. Stanton (our Editorial Project Manager) has been invaluable in producing the book and keeping things on track. In my experience, Elsevier is the best science book publisher in the world and my editorial team was the very best support that a new author could ever imagine.

I would like to thank the members of my local bike club, the Richmond Area Bicycling Association (www.RABA.org). Although we have common interests in cycling and health, RABA members are extremely diverse and individual. I wrote my parts of the book with the various RABA members in mind so that athletes of all ages and abilities would find it useful. Some of the members of the Richmond, Virginia cycling community who helped include: Jeff Nicklas, Jeanne Minnix, Beth Snyder, Bill Gary, Andrei and Alena Pugachev, Jim Strunk, Nick Morgan, Des O'Carroll, Bud Vye, Tim Mullins (www.carytownbikes.com), and RABA president Alan Cooper (who contributed to the sports photography).

My lovely wife Mary was the (sometimes reluctant?) recipient of my coaching as she improved from being a novice cyclist to becoming the female mileage leader and ride coordinator for our club. Her own initiatives in cycling and in taking up new activities such as kettlebell training helped me understand fitness and motivation from a female perspective. Mary was my general-public test reader to make sure the book was understandable and relevant to every day athletes.

Lara Crofford (an elite competitive runner) made sure that the important nutrition and physiology chapters reflected what is currently being practiced in the real world of competitive athletics, while also adding a runner's dimension to our book. Lara is a 4-time MVP of her cross-country team at the University of Nebraska, a division II track and field All-American, and a US Olympic trial qualifier in the 10,000 m. She has degrees in nutrition, health, and exercise science and is a consultant to competitive college athletes.

Dr. Edward J. Lesnefsky, Jr., M.D. took time from his busy clinical and teaching schedule to read our manuscript and offer insightful comments. Dr. Lesnefsky is the author/coauthor of over 175 journal articles, with a research focus on myocardial mitochondria at the VCU Pauley Heart Center.

Karen Smith (www.coachkarensmith.com) and her photographer Tricia Carter (www.tealtreestudios.com) supplied us with photographs of kettlebell strength training. Karen is much more than a fitness model; she is currently the only female Master StrongFirst certified instructor (www.strongfirst.com).

Kelly McQuade Kinzinger (an NSCA-CPT certified fitness trainer) reviewed the manuscript during its inception and recommends the book as a resource to fitness trainers, coaches, and her many clients. Kelly is a record-setting division I college coach, a Colonial Athletic Association "Coach of the Year," and the owner of Zinger Fit (http://ZingFitRVA.com).

My mother, Betty Hom, spent many hours proofreading my chapters. Her knowledge from working at the National Institutes of Health (NIH) as a biochemist and her proofreading skills (better than any spell check program) were invaluable. As a successful artist and art dealer, my father Jem Hom offered helpful suggestions regarding my illustrations. When he saw what I was drawing for this book, he took up painting again after a 40 year hiatus.

The Masimo Corporation (www.Masimo.com) was very generous in supplying us with samples of their new iSpO2 pulse oximetry device which we tested and reviewed. By coincidence, Masimo Corporation developed this new portable device while we were writing our book. Although I use pulse oximeters on my patients every day at the hospital, this was the first time this oxygen-sensing technology could be used by athletes in practical training.

And lastly, I would like to thank Danish fashion photographer Marc Hom (no relation). Although I have never met him, when I Google my own name the images of beautiful people fill my computer screen. You could do a whole lot worse than that.

Dr. Mark Hom, M.D.

Disclaimer

This book is not intended to diagnose or treat any disease. Although this book recommends generally accepted dietary and exercise guidelines for weight loss and disease prevention, results can vary and are not guaranteed. Although Dr. Hom is an M.D., this book is not a replacement for your seeing a doctor. This book recommends exercise (and sometimes vigorous exercise), so before starting an exercise program, consult your doctor to see if you are healthy enough to begin exercise, especially if you have any pre-existing medical conditions or are at risk for heart disease. We discourage dangerous activities, but be aware that even "safe" physical activities can lead to injuries or accidents. Before considering any of the supplements or recommended foods, consult your doctor to make sure they do not interfere with your current therapy or other medications. If you have allergies or develop an allergy to any of the foods or supplements, stop taking them immediately and see your doctor. If you think you may have a mitochondrial disease, we recommend that you speak with your physician and then consider contacting the experts at the United Mitochondrial Disease Foundation (www.UMDF.org).

Contents

Becoming a BEAST

1

The BEAST system

In the high-tech society we have constructed for ourselves, our minds are satisfied by virtual interaction in front of a computer monitor or television screen, while our bodies sit idly in a chair or repose on the couch. The Internet brings the digital world to us, instead of our need to go out and confront the real physical world. Facebook brings us virtual connections with our "friends," instead of our need to leave the comforts of home to truly personally interact. We are very adept at creating these and other false realities that satisfy our minds. However, it should be noted that the human mind is only a recent evolutionary development. The body, on the other hand, is primordial. This is the basic premise of Carl Sagan's Pulitzer Prize winning book, "The Dragons of Eden" [1] namely that animals in various forms have existed on Earth for many eons, but the human mind is merely a flash in the pan of evolutionary time. While it is true that our minds have imagined great inventions, created modern societies and shaped the world to our benefit, we have made these intellectual advancements at the detriment of our animal nature. Our physical bodies evolved under brutal conditions of death and survival. As an example of mammalian evolution, consider the wildebeest (Figure 1.1). To survive annual migration, the African wildebeest must have the endurance to stay with the herd or else be left behind. To escape a charging lion, the wildebeest must have the speed to sprint away and avoid capture. To evade a snapping crocodile, the wildebeest must have the explosive power to leap up a steep river bank.

As evolving animals we humans also had to survive as physical beings, to evade predation, to be successful predators ourselves, and to fight or flee to live another day. As a result of this process of evolution, our bodies (even now) require motion and physical activity to be healthy, vibrant, and strong. Because we no longer have to fend off lions and jackals to live in our great cities, physical activity has become less of a priority in our immediate survival. However, the diseases and conditions of the modern age and inactivity such as metabolic syndrome, type 2 diabetes, and obesity have reached epidemic proportions [2]. What is even more worrisome is that disorders such as type 2 diabetes and obesity, which were once considered adult conditions, are now common in our children [3]. The current generation of children may be the first to have a shorter life expectancy than the previous generation [4]. With less attention given to exercise and more cut backs in the physical education of our children, our long-term survival and well-being are greatly threatened. The solution (and the best way to fend off the diseases of the modern age) is to acknowledge our animalistic side and to be more physically active on a daily basis.

The Science of Fitness: Power, Performance, and Endurance. http://dx.doi.org/10.1016/B978-0-12-801023-5.00001-6

Figure 1.1 Our bodies are ancient and evolved during brutal conditions of survival and death. An example of mammalian evolution is the annual wildebeest migration in Africa. Photography: Christopher Michel.

The body–brain connection

Strengthening the body strengthens the brain. This connection between physical activity and intellectualism has been recognized by different cultures in antiquity. In ancient Greece, Plato recommended physical exercise and sport as a complement to education. In old Tibet, Buddhists considered the mind and body as inseparable, advocating physical training in preparation for higher learning.

Scientists now understand the biologic basis of the body–brain connection. Exercise boosts the brain in many ways: increasing connections between brain cells, increasing blood flow and oxygen to the brain, stimulating many areas of the brain (not just those sections involved in motor function), and increasing neurotransmitters (the vital chemicals used in nerve cell communication) [5]. There is also evidence that exercise can delay the onset and progression of neurodegenerative disorders in the brain [6]. With our aging population, exercise has never been more important in preserving cognitive function and in maintaining independence.

To an athlete, there is no greater compliment than to be called an "'animal," as it implies that barely contained within is a ferocity ready to be unleashed at any time. Eddy Merckx (the greatest cyclist of all time) was called the "Cannibal" for his hunger to win and the force with which he crushed his competition. Bernard Hinault (the great French champion) was called the "Badger" for his tenacity, toughness, and cunningness. Greg LeMond was nicknamed "Le Monster" for his dominating power and fitness.

The human body: a combination of animal and machine

Our bodies are a combination of muscle, sinew, bone, blood and hair, like any other beast. As humans we are blessed with a higher percentage of brain tissue, but under the microscope we are little different than our fellow creatures. We are animals. We are also part machine. Inside our cells, biologic processes are driven by micro "machines" such as the enzymes that perform precise and repetitive chemical reactions. Our cell energy is churned out by spinning motors inside our mitochondria [7]. Our muscles contract and move us biomechanically. Our organs work as components in concert, resulting in the human machine, which is capable of tremendous physical performance.

This book explains the science of fitness, but not in a cold laboratory setting. We will explain how to optimize fitness, the way nature intended with vigorous exercise and a healthy diet. Our methods are condensed into the BEAST system, an acronym for the five key aspects needed to optimize fitness and health. Our concept of overall health cannot be reduced further, nor are more components required. "Diet and exercise" have been the mantra of preventative health for decades, perhaps to the point of becoming cliché and disregarded. This well-intentioned advice to eat better and exercise more is not nearly specific enough to be useful. Furthermore, we think people should know *why* such recommendations are given, in order to take the advice to heart and as motivation for putting the plan into action. In this chapter we will lay out the fundamental ideas of the BEAST system and explain why each of the points fit within our entire fitness concept. In subsequent chapters we will go into much greater detail with the facts athletes need to know, such as much more specific advice about healthy eating and the best exercise methods to optimize fitness.

The **BEAST** system is composed of the following:

- Bicycling (or other aerobic exercise)
- Eating a balanced diet that supports cell function
- Avoiding toxins that impair our cells and mitochondria
- Stopping self-destructive and addictive behavior
- Training with resistance (weights)

Bicycling is recommended as a mitochondrial boosting exercise. High-intensity interval training on a bike triggers mitochondrial biogenesis (increasing the number of energy-generating dynamos inside our cells) [8]. A casual picnic ride around the lake

Figure 1.2 Single-speed commuting bike with semi-knobby tires and cut down handlebars for urban environments. Substituting bicycle commuting for car driving is one way to add more physical activity into daily life.
Photography: Mark Hom.

is not enough. The riding must be fast and strenuous. The good news is that quality (high-intensity) exercise improves fitness more efficiently than quantity (high mileage) [9]. This means that shorter but more intense workouts make the best use of the athlete's limited exercise time. Bicycling is a practical means of transportation that can replace automobile driving, proven to be deleterious to our health (Figure 1.2). The more miles we drive in a car, the more overweight and unhealthy we become [10]. Of course, there are alternative means of aerobic exercises such as running, swimming, triathlons, rowing, cross-country skiing, and gym workouts. They can be just as good as long as they are practiced with high intensity. The other factor in the exercise equation is frequency, as infrequent exercise and sloth result in mitochondrial atrophy. Fitness is truly a case of use it or lose it.

Eating right is often taken for granted, misunderstood, or ignored. Think about eating for the future and not for immediate gratification. What we eat (good or bad) eventually becomes incorporated in our organs, cells, and mitochondria. There are recommended foods and nutrients that support exercise and mitochondrial health. It is not as simple as saying that carbohydrates are bad (they rebuild glycogen stores), that all fats are bad (certain types of fat are beneficial), or that all proteins are good (processed meat can contain unhealthy saturated fat and additives). There are several pro-mitochondrial nutrients that support athletic energy production. Just be aware that there is no such thing as exercise in a bottle. Eating right means plenty of fruits and vegetables (rich in vitamins and antioxidants), non-animal sources of protein, quality red meat (in sparing amounts), healthy fats for strong membranes, timely intake of carbohydrates to fuel exercise and to rebuild energy reserves, and portion control to match exercise output. Flavor and variety are not sacrificed.

Avoiding toxicity is a key part of fitness because our cells and mitochondria are sensitive to a wide array of toxic substances. Although it was once wrongly assumed that mitochondria were perfect entities that never malfunction, we now know that mitochondria are vulnerable and therefore we must take good care them. Out of convenience and productivity, our food supply is infused with a multitude of unnatural chemicals and additives. One example is trans fat, which is artificially altered to extend the shelf life of convenience foods, but degrades the membranes of our cells and mitochondria [11]. Strong membranes are key components in mitochondrial energy production and in protecting the cell from free radical damage. Certain commonly prescribed medications can be harmful to mitochondria and should be avoided when possible. Exposure to heavy metals, pesticides, and other environmental toxins must be avoided.

Stopping self-destructive behavior is not as easy as it sounds. For example, nicotine is one of the most highly addictive substances known to man. Heroin addicts often say it is easier to quit heroin than to quit smoking [12]. Cigarette smoke is highly damaging to the lungs, heart, and arteries (at the organ level), and poisonous to the mitochondria (at the sub-cellular level) [13]. Alcoholism can damage the liver, a major organ of energy metabolism. Overeating can lead to metabolic syndrome, type 2 diabetes, and obesity. Exercise addiction can lead to over-training, over-use injuries, and often decreased performance. Although all of these addictions are difficult to overcome, they are easier to conquer when we exercise and begin to respect and appreciate our bodies. While we do not want to preach, we realize that we are not always perfect and must recognize and correct self-destructive behaviors if we want to improve.

Training with resistance (as a complement to aerobic exercise) increases lean muscle mass, raises metabolism, sculpts the body, improves posture, and strengthens the bones [14]. Strength training can take the form of circuit training at a gym, free weights, home exercise equipment, push-ups and pull-ups, Pilates, or kettlebell training (Figure 1.3). Elite cyclists and runners have very specific body shapes that may be optimal for their respective sports, but for most of us a more balanced approach is better for overall health. Science has brought us a better understanding of the benefits of strength training: why working against gravity is important for health (NASA microgravity studies) [15]; how specific exercises target the different types of muscle fibers; the mechanisms of muscle endurance and fatigue; muscle energetics and recovery; how core strength supports power output; the connection between strength training and bone health; and the importance of muscle mass in metabolism.

What is an athlete?

Athlete: a person possessing the natural or acquired traits of strength, agility, and endurance that are necessary for physical exercise or sports.

"Give me a word, any word, and I show you that the root of that word is Greek."
Gus Portokalos from the movie "My Big Fat Greek Wedding" [16].

"Now let us go outside and try our skill in various sports, so this stranger when he is home can tell his friends how much better we are than other men in boxing, wrestling, running, and leaping."
Homer from "The Odyssey" [17]. Ancient sports trash talking.

The word "athlete" comes from the Greek root word "athlos" (contest). The Greeks were the first western culture to recognize the importance of athletic competition. Homer was one of the first to describe the drama of physical contests in his epics the Iliad and the Odyssey. The ancient Olympic Games were held in honor of the King of the Gods, Zeus, and it was assumed that the competitions took place under the watchful eye of Zeus himself. Olympic spirit includes the concepts of fair play and ethical sportsmanship, often forgotten in the modern era of "win at all costs."

For the purposes of "SciFit," we define the athlete as anyone who engages in physical activity for the purpose of improving themselves. We contend that motivation for physical exercise should be fitness improvement. This core principle applies to athletes of all abilities and fitness levels, competitive or non-competitive. It does not matter whether you are an Olympian, Tour contender, or someone with a new trial membership at the gym. If your goal is to become more fit, this book will point the way. Indeed, the authors of this book came to athletics and fitness from very different perspectives: one seeking victory at the very highest level of sports competition, and the other seeking improved vitality and health by investigating the science of exercise physiology.

We want every athlete – from recreational to professional – to have the advantage of knowing what we have spent years learning: the scientific basis of physical training and what it takes to reach ever higher levels of fitness. The secret to becoming stronger and more physically fit lies within our cells, and in the tiny yet powerful structures called mitochondria. This is how every athlete can reach their goals and become scientifically and physically fit.

The name BEAST is to remind us that we are animals that require frequent physical exercise to maintain health. Although the modern mind can be satisfied with the virtual world of the Internet and Facebook, the body is ancient and evolved in a physical realm of sweating, grunting, and panting. When we take care of our bodies, we acknowledge the beast within us. When you become a beast, you will become stronger and healthier (Figure 1.4).

Figure 1.3 Training with weights (in this case with kettlebells) complements aerobic exercise for total body fitness.
Model: Karen Smith (Master StrongFirst Instructor). Photography: Tricia Carter.

Figure 1.4 2007 U.S. Open Cycling Championships in Richmond, Virginia.
Photography: Mark Hom.

References

[1]   Sagan C. The Dragons of Eden: Speculations on the Evolution of Human Intelligence. New York: Ballantine Books; 1977.

[2]   Mensah GA, Mokdad AH, Ford E, et al. Obesity, metabolic syndrome, and type 2 diabetes: emerging epidemics and their cardiovascular implications. Cardiology Clinics 2004;22(4):485–504.

[3]   James PT, Rigby N, Leach R. International Obesity Task Force. The obesity epidemic, metabolic syndrome and future prevention strategies. European Journal of Cardiovascular Prevention & Rehabilitation 2004;11(1):3–8.

[4]   Carmona RH. Surgeon General. The Growing Epidemic of Childhood Obesity. U.S. Public Health Service, U.S. Department of Health and Human Services; 2004.

[5]   Ratey JJ, Hageman E. Spark: The Revolutionary New Science of Exercise and the Brain. New York: Little Brown and Company; 2008.

[6]   Ang E, Tai Y, Lo S, et al. Neurodegenerative diseases: exercising toward neurogenesis and neuroregeneration. Frontiers in Aging Neuroscience 2010;2:25.

[7]   Boyer PD. The ATP synthase – a splendid molecular machine. Annual Review in Biochemistry 1997;66:717–49.

[8]   Little JP, Safdar A, Bishop D, et al. An acute bout of high-intensity interval training increases the nuclear abundance of PGC-1α and activates mitochondrial biogenesis in human skeletal muscle. American Journal of Physiology - Regulatory, Integrative and Comparative Physiology 2011;300(6):1303–10.

[9]   Little JP, Safdar A, Wilkin GP, et al. A practical model of low-volume high-intensity interval training induces mitochondrial biogenesis in human skeletal muscle: potential mechanisms. Journal of Physiology 2010;588(6):1011–22.

[10] Hoehner CM, Barlow CE, Allen P, et al. Commuting distance, cardiorespiratory fitness, and metabolic risk. American Journal of Preventative Medicine 2012;42(6):571–8.

[11] Ibrahim A, Natarajan S. Dietary trans-fatty acids alter adipocyte plasma membrane fatty acid composition and insulin sensitivity in rats. Metabolism - Clinical and Experimental 2005;54(2):240–6.

[12] Blakeslee S. Nicotine: harder to kick...than heroin. The New York Times, March 29, 1987.

[13] Masayesva BG, Mambo E, Taylor RJ, et al. Mitochondrial DNA content increase in response to cigarette smoking. Cancer Epidemiology, Biomarkers & Prevention 2006;15(1):19–24.

[14] Seguin R, Nelson ME. The benefits of strength training for older adults. American Journal of Preventative Medicine 2003;3(2):141–9.

[15] Bagley JR, Murach KA, Trappe SW. Microgravity-induced fiber type shift in human skeletal muscle. Gravitational and Space Biology 2012;26(1):34–40.

[16] Vardalos N. My Big Fat Greek Wedding. IFC Films; 2002.

[17] Homer. The Odyssey. Various translations. Circa 8th century BC.

The Human Machine

2

The human body as a machine

Most fitness books begin and end with muscles. We too will describe muscles in depth because they are responsible for athletic speed, power, and endurance. Athlete performance is a result of muscle power. Winning is a result of applying that muscle power at the right time. Muscles are what you feel "burning" when pushed into the anaerobic (low oxygen) zone. Muscles are what ache after a strenuous workout. Muscles are what hypertrophy (enlarge) when you train with resistance. It is easy to fall into the trap that all you need to think about are your muscles (Figure 2.1). In our broader and more complete explanation of fitness, the first major concept is that fitness is a whole body process that requires many interacting systems working in concert. Only when these systems are in harmony can true fitness and maximum performance be attained. This book is different because it explains all of the organ systems involved in athletic performance, not just muscles. Each organ system has a specialized role in energy metabolism. Understanding how the human body works is the first step in optimizing your fitness. An anatomist dissecting a dead creature will separate the individual organs from each other, subdivide structures within each organ, all the while applying names and subclassification systems. For example, the right lung in humans is divided into an upper lobe, middle lobe, and lower lobe. Within each lobe are named segments, bronchial airways, arteries, veins, and air sacs. It is easy to fall into another trap of believing that subclassification by itself is all one needs to know about a subject. In a living and breathing person, it is more important to know the physiologic function and the interaction between multiple systems. In order to explain how the body works, we will have to bend some rules. For example, body fat (adipose tissue) is usually considered part of the organ system of the skin. Although subcutaneous fat (just under the surface layer of the skin) does serve as a protective and insulating layer, there is also fat deep inside the body, surrounding our internal organs. We also now know that fat serves many key functions in athletic metabolism. We therefore describe body fat as one of the organ systems of athletic performance, even though an anatomist would be upset at us giving it too much importance.

The major organ systems of athletic performance include:

1. *The Digestive and Endocrine System:* The process of transforming bulk food into the simple molecules we can absorb into our blood stream and into our cells. This section will explain the differences in how carbohydrates, protein, and fat are digested, why carbohydrates are the most easily digested fuel, how sports drinks improve performance, how carbohydrates are stored in the body, how glucose is tightly regulated in the blood stream (i.e., how insulin works), why athletes need protein even though it takes longer to digest, and why dietary fat is important and how it satisfies hunger. Sports nutrition (a multi-billion dollar industry) will be explained at the cellular level in Chapter 4: Feeding Your Cells and Chapter 5: Mitochondrial Supplements.

The Science of Fitness: Power, Performance, and Endurance. http://dx.doi.org/10.1016/B978-0-12-801023-5.00002-8

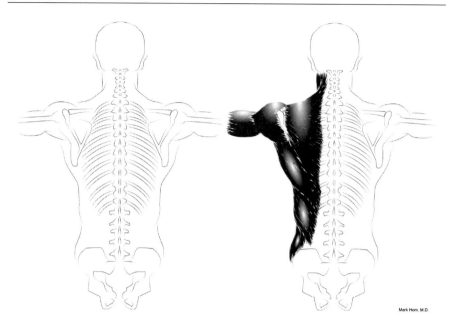

Mark Hom, M.D.

Figure 2.1 Most other fitness books begin and end with muscles. In the SciFit concept, muscles are only one part of the human machine.
Image by Mark Hom.

2. *The Liver:* The under-looked and under-appreciated organ of athletic metabolism. Despite being the largest internal organ, the liver's key role in athletic metabolism is not understood by most athletes. The liver stores glycogen, releases glucose, switches between metabolic states, clears lactic acid, detoxifies the body, produces cholesterol (both good and bad forms), and is the body's chemical factory. Although the liver does not move or pulsate, it is highly metabolically active and is loaded with mitochondria. Most fitness books ignore this vital organ, but the liver is the hub of metabolism and assists the athlete when recovering from intense exertion. For effective attack and counter attack during a race, the liver must be healthy. Alcoholism, obesity, hepatotoxic medications, and blood borne infections (viral hepatitis) can impair liver function.

3. *Body Fat (Adipose Tissue):* The body stores energy in its most concentrated form as body fat. Although fitness athletes abhor body fat, without energy-dense fat we would only store energy as glucose and glycogen (much greater in volume per calorie) and we would all be truly enormous. Besides energy storage, body fat serves as insulation and protective cushioning, regulates body temperature, keeps us alive during starvation, and is distributed throughout our bodies in varying ways depending on our sex, age, and metabolism. Therefore due to its mass and unique importance in athletic energy, adipose tissue is considered here as a separate organ. Nearly all fat burning takes place inside mitochondria. Proper training burns body fat and can increase power to weight ratio, but it is not healthy (nor possible) to eliminate body fat entirely. The endurance athlete relies on adipose tissue for a steady supply of fuel, and also as reserve energy for staying power and second wind. By training for endurance, the athlete will burn fat and spare the body's much more limited supply of glycogen.

4. *The Heart and Oxygen Delivery*: To generate the energy we require, our mitochondria require two things: oxygen and fuel. The heart beats continuously, pumping oxygen and fuel to the organs and muscles. The heart is connected (subconsciously) to our brain and nervous system, responding immediately when we pick up the pace when exercising. The heart is the first organ to receive oxygenated blood (via the coronary arteries) and heart muscle tissue has the highest mitochondrial density in the body. Oxygen delivery to muscles is optimized with a strong heart, clear lungs, open arteries, proper muscle tension and relaxation, and a healthy number of red blood cells. Cardiac output (powered by mitochondria) is a major determinant of fitness improvement and VO_2 max. Blood cell production (hematopoiesis) is a very complicated process requiring a long list of vitamins and minerals (not just iron). This is why athletes should take a multivitamin and why vegetarian athletes need to take extra care of their nutrition. Cardiovascular disease is the leading cause of death worldwide and weak heart muscle is the cause of much disability. Cigarette smoke and air pollution can permanently damage the delicate alveoli (air sacs) in our lungs.

5. *Skeletal Muscle*: An athlete's muscles cannot burn glucose or fat directly; these fuels must be converted by mitochondria into adenosine triphosphate (ATP, the energized molecule that powers muscle). Contracting muscle fibers consume many trillions of molecules of ATP during exercise and therefore boosting mitochondria is the sure way to maximize performance. There are different types of skeletal muscle fibers, notably different in their mitochondrial density. This allows the body to perform with both powerful bursts and sustained effort. Anaerobic strength, aerobic endurance, speed, power, and fatigue will be explained in this section. The practical and performance aspects of muscles will be detailed in Chapter 6: Your Body at Work, Play, and Rest, Chapter 10: Gauging Fitness, and Chapter 11: BEAST Fitness Training.

6. *Homeostasis (Fluid and Temperature Balance)*: The body functions within fairly narrow parameters, regulated by multiple organ systems. Mitochondria not only supply the chemical energy to these systems, but also generate heat and water as by-products. The athlete must prevent (and be cognizant of the symptoms of) dehydration, salt imbalance, heat stroke, and hypothermia. By understanding homeostasis, the informed athlete will perform better and perform safely under all conditions.

7. *The Immune System*: The body needs a strong defense: healthy skin (the first barrier to infection), good teeth to lessen bacterial exposure, "good" bacteria in the colon (to suppress harmful bacteria), adequate rest and recovery, and strong white blood cells (powered by mitochondria). Infection, chronic inflammation, autoimmune disorders, over-training, vitamin deficiencies, and weakened immunity can impair athletic performance. The emergence of antibiotic-resistant bacteria is becoming an increasingly dangerous threat to public health.

8. *The Brain and Nervous System*: Often called the selfish organ, the brain makes sure its energy needs are met before all others. Long chain fatty acids cannot cross the blood brain barrier, so the brain largely depends on carbohydrates (sugar and a limited supply of stored glycogen). The brain's sugar craving contributes to obesity in sedentary people. The athlete can eat lots of carbohydrates yet burn them off, keeping the mind and body in better harmony. The brain re-accumulates its glycogen at night, a re-energizing process which explains the mystery of why we require sleep. Athletic performance can only be optimized when the brain is satisfied. The brain is the master control organ and is hard wired to the other organs via the peripheral and autonomic nervous systems. Skeletal muscle is controlled voluntarily by the cerebrum and coordinated by the cerebellum. Many other systems are controlled autonomically (without conscious thought).

Digestive and endocrine systems

Digestion is the process that converts a granola bar into the sugar used for energy, and the process that converts a beef steak into the amino acids that grow bigger muscles. Anatomically it is often called the gastrointestinal tract but digestion is more than just the stomach and bowels. The digestive process begins in the mouth and along the way is assisted by many other organs such as the liver (bile), pancreas (digestive enzymes), circulatory system, and immune system.

The lowly sea cucumber

A good demonstration of the digestive system is the sea cucumber. Despite its vegetable name, it is actually an animal that crawls and burrows along the bottom of our oceans. Its elongated body shape is merely a housing for its digestive tract. One end is its mouth, lined with frond-like appendages that collect and stuff food into its oral cavity. The lowly sea cucumber does not have a brain, rather a ring of nerve cells around its mouth, which tell it when to feed, crawl, hide or spawn. Its digestive tract separates the sand from its food and allows it to absorb nutrients from the detritus on the sea bottom. The other end is its anus, which it uses to breath, defecate, and defend itself (with retractable tentacles). The sea cucumber is a simple animal that looks like and acts like a crawling piece of intestine. Although most people consider it disgusting, a biologist sees the sea cucumber as a great success. With its simple means of existence, it has survived the eons without requiring significant change. Fossils of sea cucumbers date back 450,000,000 years [1] and they continue to thrive today, pointing to the importance of the digestive system. Most people consider dinosaurs to be evolutionary failures, yet dinosaurs were the dominant animals on Earth for 135,000,000 years [2]. Humans (*Homo sapiens*) have only been around for 195,000 years [3] and we already have created the means to annihilate ourselves with biologic, chemical, and nuclear warfare. So do not pity the lowly sea cucumber. Its simple yet effective body (optimized for digestion), will surely last longer than us.

The adult human digestive tract is about 30 feet in length. The word "about" is used because the digestive tract is continually contracting to squeeze food on its way (peristalsis). This contracting and squeezing is performed by involuntary smooth muscle and requires energy and oxygen. When athletes begin exercising, blood is shunted more towards the skeletal muscles and less to the digestive tract [4]. It is common for beginning athletes to have abdominal cramps and indigestion, which can improve with training and well-timed nutrition, but can worsen with extreme exercise duration and dehydration [5]. Before exercise and during exercise it is important that the athlete eat food that is easy to digest and requires less intestinal blood flow. Furthermore, the digestive tract requires a large amount of fluid to keep its insides flowing and lubricated. The salivary glands produce up to 1.5 L of saliva per day and the stomach produces a similar volume of gastric juices [6]. The rapid breathing during exercise dries the mouth with evaporation (try blowing on a cool mirror) and perspiration through the skin causes loss of water and salts. These fluid losses and the need to keep blood flowing to both the muscles and intestines are reasons to stay well-hydrated during exercise.

The easiest foods to digest are carbohydrates: sugar, starches, and grains. That is because digestion of carbohydrates begins right away, when food is still being chewed in the mouth. Saliva contains mainly water, but also lubricating mucous, antibacterial IgA and lysozymes, and digestive enzymes [7]. One enzyme, amylase, converts the starch in bread into sugar. If you keep a piece of bread in your mouth for a short while it will begin to taste sweet. Table sugar stimulates the taste buds on the tongue and gives the pleasurable sensation of sweetness. Sugar is further broken down into simple sugars such as glucose and fructose (by the enzyme sucrase) and is rapidly absorbed into the blood stream along the remainder of the gastrointestinal tract. Note that enzymes have the "-ase" suffix). Glucose is hydrophilic (water attractive) so when one molecule of glucose is absorbed, 6 molecules of water come along with it. One benefit of sports drinks is that the sugars and sodium they contain hasten the absorption of water and thus you can rehydrate with a sports drink faster than with plain water [8]. Energy "gels" use the principle of rapid absorption by containing concentrated sugars as a means of quick refueling with less strain on the digestive system. However, be aware that excess calories of any form can be converted into body fat, and that spikes in blood glucose cause a strain on the blood sugar regulatory system (i.e., insulin). Instead of exercising into carbohydrate depletion, a better plan is to eat whole grains which slow and moderate the absorption of carbohydrates, allowing for a more constant source of fuel and less extreme spiking of blood glucose and insulin. The correct amount of post exercise carbohydrates will top off glycogen stores, yet not cause increase in body fat.

What is sugar?

Table sugar (sucrose) is a disaccharide consisting of one molecule of glucose and one molecule of fructose. Glucose and fructose are monosaccharides (also known as simple sugars). Starch is many molecules of glucose bound together (polymerized) found in grains and potatoes, and glycogen is the polymerized form of glucose found in animals. Simple sugars, table sugar, starch, and glycogen are all carbohydrates in various forms. Table sugar stimulates the taste buds on the surface of the tongue, giving us the pleasurable sensation of sweetness. Fructose and sugar substitutes are actually sweeter than table sugar and glucose. The terms blood sugar and blood glucose are often used synonymously. The glycemic index refers to how rapidly carbohydrate foods raise blood glucose levels which is important for controlling diabetes. Foods with low glycemic index include beans, nuts, vegetables, and fresh fruit. Foods with a medium glycemic index include whole grains, dried fruit, and table sugar. Table sugar is only half glucose and the fructose half is slowly converted to glucose by the liver. Foods with a high glycemic index include processed baked goods, white rice, and glucose. Although there is some concern about High Fructose Corn Syrup (HFCS), table sugar is 50% fructose and HFCS is 55% fructose [9]. Fructose is sweeter than table sugar. The problem is that HFCS is inexpensive, easy to add to processed food, bare of other nutrients or fiber, and in too high of a dose in some processed foods. As with all food, moderation is the key.

Blood glucose levels are controlled by the pancreas via two hormones: insulin and glucagon. When a sugary or starchy meal is eaten, glucose is absorbed by the digestive tract and blood glucose levels rise (how much depends on the quantity of the meal and its glycemic index). The pancreas responds by releasing insulin into the blood stream which is the signal to take up glucose for storage inside of the cells and organs, thus bringing blood glucose levels back down to normal. The glucose inside of cells can be stored as glycogen (many molecules of glucose polymerized into spherical granules), most notably in certain organs such as muscles and the liver. In type 1 diabetes (once called juvenile onset diabetes), the pancreas does not make enough insulin and blood glucose levels remain too high. In type 2 diabetes (once called adult onset diabetes) the cells become resistant to the insulin signal and blood glucose levels remain too high. Short-term high blood glucose levels do not cause symptoms, but in both forms of diabetes long-term high blood glucose levels allow glucose to cling onto other vital chemicals in the body, impairing their function and leading to problems with blood vessels, oxygenation, vision, nerves, and organs. An example is when glucose binds to red blood cells (making them less pliable) and hemoglobin (making it harder to release oxygen to tissues). Hemoglobin A1c (HbA1c) is the sugar coated form of hemoglobin and can be measured in a blood test to determine long-term high blood glucose levels (over the last 90–120 days depending on how fast the red blood cells are replaced) [10]. Insulin also triggers the conversion of glucose into fat in body fat cells (adipose tissue). Insulin also triggers the uptake of amino acids and fats into cells. When blood sugar levels are too low, the pancreas releases the hormone glucagon, which has the opposite effect of insulin. Glucagon signals the liver to break down stored glycogen and release glucose (Figure 2.2).

Protein in the form of meat takes longer to digest and requires more blood flow to the gastro-intestinal tract. Red meat is nutrient dense with protein of course, but also many essential vitamins and minerals such as iron [11]. Unlike sugar which is a simple molecule, meat is a very complex structure that must be chewed into small pieces, dissolved in stomach acid, and digested by many enzymes. The building blocks of protein (amino acids) are what the intestines can absorb and what builds new muscle. Red meat is nutritious, but also comes with saturated fat and in some cases steroids, growth hormone, and antibiotics [12]. Red meat has benefits such as a form of iron than can be absorbed better than plant or pill sources of iron [13], but red meat does not have to be eaten every day (Figure 2.3). In the past, football players were often given a steak breakfast, perhaps for psychological reasons. But meat is harder to digest and takes blood away from working muscles. A better pregame food would be a whole grain cereal or oatmeal to supply energy to muscles. For post-event recovery, protein shakes with easier to digest powdered protein from plant and animal sources and added carbohydrate will assist the rebuilding of muscle and glycogen stores.

Fat digestion is also important and greatly misunderstood. Not all dietary fat is bad. In fact there are many dietary fats that are beneficial to health and performance. Studies have shown that the omega-3 oils in a high-fat fish diet protect against cardiovascular disease [14] and the monounsaturated fat in olive oil is one reason why the Mediterranean diet is called the healthiest diet in the world [15]. When you eat a fatty meal, the pancreas releases digestive enzymes (lipases) and the gallbladder contracts and releases

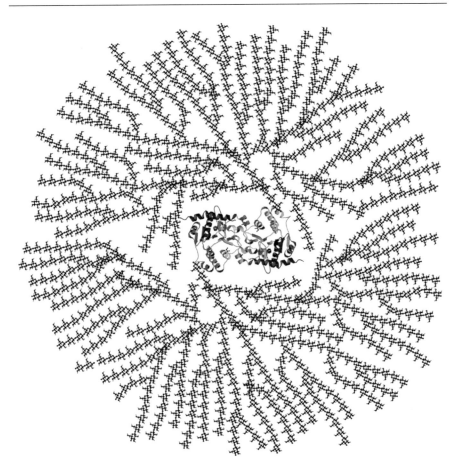

Figure 2.2 A glycogen granule made of polymerized glucose.
Image by Mikael Haggstrom.

bile into the duodenum. The bile forms micelles which are tiny fat droplets that the intestines can absorb. This is also important in the absorption of fat soluble vitamins (A, D, E, and K). Hunger, fullness, and satiety are complex feedback sensations that determine how much and how often we eat. One component is the effect of dietary fat in satisfying hunger and telling us when we have had enough [16]. Blood levels of glucose, amino acids, and fatty acids trigger hormones that shut off hunger sensations. Gastric distention is also a factor and bulkier foods containing fiber make us feel full. Fiber also lowers glycemic index, decreases saturated fat absorption, and lowers cholesterol, which are all good for health [17]. By helping remove toxins and harmful bacterial from the intestines, fiber is known to lower the risk of colon cancer [18]. The typical modern diet is far too low in fiber, due to overly processed and purified ingredients. While it is true that a high fiber diet increases bulk and creates more work for your intestines, you can gradually increase the fiber content in your diet and avoid symptoms of bloating. Adequate fiber and good hydration can prevent irregularity and constipation.

Figure 2.3 Red meat has protein, vitamins, and iron. Drawing based on a photograph from the Food Channel (www.foodchannel.com). Image by Mark Hom.

The liver

The liver has some very unique features and many metabolic functions. As the largest internal organ, it occupies the right upper quadrant of the abdomen. Although it does not beat like the heart or billow like the lungs, the liver is a vital organ and we can only live about 8 hours without it. It has many functions including carbohydrate metabolism, fat metabolism, cholesterol production, blood detoxification, bile production, digestion, and lactic acid clearance. There is an old saying that life depends on the liver, which is true. When liver is diseased, patients suffer with jaundice (yellowing of the skin and eyes), pruritus (itching of the skin), ascites (water in the abdomen), internal venous backups, and in the late stages: hepatic encephalopathy and coma. The ancient Greeks noticed that the liver could regenerate itself. In the Myth of Prometheus, as punishment for stealing fire from the Gods and giving it to mankind, Prometheus was tortured by having his liver was eaten away by an eagle, only have it grow back the next day to be eaten again. Liver regeneration can be advantageous in the case of trauma to the liver, as it can heal itself. However, in the case of chronic viral hepatitis this constant regrowth leads to cirrhosis (scarring and nodularity) with a risk for hepatocellular carcinoma (liver cancer) [19]. The liver has a dual blood supply. Besides an arterial supply that all other organs have, it also has a venous blood supply from the portal vein. The portal vein drains the blood directly from our intestines and therefore our liver has the first shot at the nutrients we absorb from digestion. The liver makes cholesterol. If you listen to advertisements on television you would think that cholesterol is some sort of poison that can be avoided in food and can be removed with pills (statin medications). In fact cholesterol is a vital part of our cell membranes and several key hormones. It should also be noted that very little dietary cholesterol is absorbed from food [20]. What is important to cardiovascular health is the balance and form of cholesterol. low-density lipoprotein (LDL, known as "bad cholesterol") delivers cholesterol from the liver to the rest of the body, where in excess it can cause plaques and arterial narrowing. On the other hand, high-density lipoprotein (HDL, known as "good cholesterol") removes cholesterol from the body and brings it back to the liver [21]. The liver uses this cholesterol to make bile that aids in the digestion of fat. The bile gets reabsorbed and recycled by the intestines. Dietary fiber soaks up and eliminates bile (as well as some dietary fat) from the system and breaks the recycling of bile. To compensate and to make more bile, the liver draws more cholesterol from the blood stream thus lowering cholesterol in the body. A typical modern diet is very low in dietary fiber due to overly processed food, which concentrates the caloric component of food and removes the less digestible fiber component. This is why whole grains and supplemental dietary fiber are needed to lower cholesterol and improve health [22].

The liver can switch between several metabolic states. When we eat a meal and nutrients enter the blood stream, insulin from the pancreas signals the liver to go into energy storage mode, taking glucose from the blood and storing it as glycogen in hepatocytes (liver cells). When we exercise or when blood glucose levels drop, glucagon from the pancreas signals the liver to energy release mode, freeing stored glycogen in the form of blood glucose. With maximal or near maximal exercise, the muscles go into anaerobic (low oxygen) mode, which is less energy efficient and creates lactic acid. Unless eliminated by the kidneys or converted by the liver, lactic acid can build up in the blood stream and impair performance and in the extreme case cause lactic acidosis (acidification of the blood). The liver can convert lactic acid back into glucose. Although it costs more net energy for the liver to perform this feat, it clears lactic acid from the blood, turns a byproduct into a fuel, and enables the athlete to recover. Most people think that only plants can make sugar, but your liver can make sugar too. The process is called gluconeogenesis and the interplay between muscle and liver with glucose and lactic acid is called the Cori cycle [23]. In times of starvation, glycogen stores become depleted and the body must start turning body fat into fuel. The liver responds by chopping the long chains that make fatty acids into shorter pieces called ketone bodies. This process of ketosis developed as a means of maintaining energy and surviving during times of famine during our evolution. Ketones can be measured in the blood, urine, and breath (acetone). Unlike long chain fatty acids, ketones can cross the blood brain barrier and supply the brain in times of starvation [24]. Although starvation will cause weight loss, it is not a healthy way to lose fat. Starvation triggers the release of stress hormones that encourage the rebuilding of body fat once the starvation ends. The empowerment some feel from losing weight can trigger the abnormal psychology and distorted body image of anorexia nervosa. In prolonged starvation, the body will also begin braking down the protein in muscle into amino acid fuel, so the muscle mass that took so long to build will quickly melt away.

Adipose tissue (body fat)

Body fat was once considered just a means of caloric energy storage and the bane of overweight athletes. But recent studies have shown that body fat is much more complex and dynamic having key roles in basal metabolism, temperature regulation, energy reserves for endurance athletes, and as the main fuel for the heart [25]. The type and distribution of our body fat is a result of our genes, different in men and women, changes with age, and is predictive of health risks. A common way to measure body fat is body mass index (BMI), which is a calculation using height and body weight for comparison on standardized charts. Although BMI is easy to measure, track, and compare, it has several limitations [26]. It does not directly measure percent body fat and assumes that excess body weight is fat and not muscle (body builders can be wrongly classified as obese). Also as we get older we tend to lose muscle. If you lose muscle and replace it with body fat your body weight may not change and your BMI can stay the same although your body composition is different. BMI also does not take into account where the fat is distributed in the body. Although most people think about the subcutaneous fat just under the skin surface, there is also fat deep in the body surrounding our abdominal organs. This type of fat is correlated with health problems. The typical male distribution of deep abdominal fat (beer belly) is more dangerous to health than the typical female distribution of subcutaneous fat in the hip and thigh regions [27].

Interpretation of BMI for adults: For adults 20 years old and older, BMI is interpreted using standard weight status categories that are the same for all ages and for both men and women (Table 2.1). For children and teens, on the other hand, the interpretation of BMI is both age and sex specific. The standard weight status categories associated with BMI ranges for adults are shown in Table 2.2.

For medical students, their first exposure to body fat comes from first semester anatomy class when cadavers are dissected. Body fat is a gelatinous slimy substance that is nauseating to touch and smell and gets in the way of the more interesting organs. This begins young doctors' bias against overweight patients. It is easy to assume that an overweight person chooses to be unhealthy. However, obesity is a complex issue. Some forms of obesity are due to hormone imbalance (such as Cushing's syndrome from too much cortisol). Obesity by itself is not a direct cause of cardiovascular

Table 2.1 Body Mass Index (BMI) calculation

Kilograms and meters	BMI = weight in kg/(height in meters)^2 Example: weight = 68 kg, height = 165 cm = 1.65 m Calculation: $68 \div (1.65)^2 = 24.98$
Pounds and inches (note: conversion factor)	BMI = [703 × weight in lbs]/(height in inches)^2 Example: weight = 150 lbs, height = 5′5″ = 65″ Calculation: $[703 \times 150] \div (65)^2 = 24.96$

Table 2.2 Body Mass Index (BMI) ranges

BMI	Weight status
Below 18.5	Underweight
18.5–24.9	Normal
25.0–29.9	Overweight
30.0 and above	Obese

disease, although the same lifestyle choices can increase the propensity for both. Obesity is ironically associated with poverty, because inexpensive food is often highly processed to be high in calories and low in nutrients.

As noted by Glenn A. Gaesser [28]: one's fitness level is a better determinant of health (lower mortality) than being overweight (less effect on mortality); thinness can have a negative effect on health (increasing mortality); and when an overweight person begins to exercise, there are measurable improvements and health benefits, independent of weight loss.

We are also discovering that body fat has key roles in metabolism. For example, insulin is best known for lowering blood glucose levels, but insulin also removes lipids from the blood and increases fat storage. The heart muscle derives most of its fuel from body fat, which is a more steady and reliable source compared to sugar. Fat is the friend of the endurance athlete because it is the source of sustainable energy and the reserve power athletes feel when they get their second wind. As we will describe in subsequent chapters, fat burning takes place in our mitochondria, so exercise is the best way to keep body fat in proper balance and proportion.

Although it was once thought that brown fat (brown adipose tissue) was only present in babies, small rodents, and hibernating bears, recent discoveries in positron emission tomography (PET scanning) have proven the existence of metabolically significant amounts of brown adipose tissue in adults. PET scanning uses technology straight out of a "Star Trek" movie. A cyclotron generates radionuclides that emit positrons (the antimatter to electrons). The radionuclides are bound to sugar molecules that are then injected into the blood stream and go to metabolically active tissues such as the brain, heart, and tumors. When a positron meets an electron they annihilate each other in a flash of light traveling in opposite directions. This light can be measured and located in the body giving us a three-dimensional image of the body. When PET scanning was being developed, researchers noted activity at the base of the neck and along the shoulder blades in some healthy adult test subjects. Biopsies revealed that this activity was due to brown fat [29]. Recent research has demonstrated the importance of brown adipose tissue in temperature regulation, caloric expenditure, basal metabolism, and in predicting obesity risk. The more obese a person is, the less brown fat they tend to have [30]. Brown fat is brown because of the heme iron in mitochondria and the high mitochondrial density in brown fat. Mitochondria have an internal proton gradient (a type of battery) that powers the generation of ATP. Brown fat mitochondria can uncouple the proton gradient to generate pure body heat instead of ATP. Brown fat is important to maintaining normal body temperature in cold environments and also is a major component of body temperature regulation and basal metabolism. Therefore we include brown adipose tissue not just as an aside, but as another example of how mitochondrial fat burning is a key component of our metabolism.

Excess body fat does have negative health consequences. It can increase insulin resistance and lead to metabolic syndrome and type 2 diabetes as we will describe in Chapter 8: When Things Go Wrong. Obesity can lead to high blood pressure, arthritis, disability, and low self-esteem. Mitochondria are the center of fat burning and therefore the key in understanding and controlling obesity. The goal of this book is not to judge, but to give everyone the opportunity to maximize their physical potential and improve their health.

The heart and oxygen delivery

The ultimate power molecule in the body is ATP (adenosine triphosphate). Every heartbeat, foot step, and mental process requires ATP energy. How efficiently you generate and use this ATP determines how fast, powerful, and enduring you can be as an athlete. When you exercise in aerobic (oxygen rich) conditions, your mitochondria can generate 34 ATP per glucose. When you exercise in anaerobic (low oxygen) conditions, your cells use less efficient glycolysis, which generates only two ATP per glucose. Therefore when your cells have enough oxygen, they are 17 times more energy efficient. When your mitochondria have fuel and enough oxygen to burn that fuel, you will perform at a faster pace, sustain that pace longer, and recover faster. Modern athletes are much better informed about nutrition and know that they need carbohydrates to fuel their muscles, and because our muscles and liver can store energy as glycogen and fat, food is usually not the rate-limiting factor in performance (at least for short and medium duration exercise). With proper physical training an athlete builds more muscle and more mitochondria and therefore has increased oxygen demand. This is why improving oxygen delivery to the muscles can make a dramatic difference in physical performance. The oxygen delivery system in your body consists of the heart (pumping), lungs (gas exchange), and the blood and blood vessels (transport). In the short term this oxygen delivery system makes rapid and dramatic changes from rest to all-out effort. In the long term the heart adapts to training and becomes stronger as a pump. Strengthening your heart and strengthening your skeletal muscles are the basis of fitness improvement. Both your heart and your muscles are powered by mitochondria.

Of all the organs and tissues in the body, the heart is the most metabolically active and has the highest concentration of mitochondria. In a healthy person, mitochondria take up 1/3 of the volume of heart muscle. The heart is composed of striated muscle as well as valves, arteries, veins, and nerves. The heart works at all times and cannot take a day (or minute) off the job. When we exercise, our muscles demand more oxygen and a healthy heart responds by pumping faster (increasing heart rate) and pumping more forcefully (increasing stroke volume). The increase in heart rate is easier of the two to measure, either with a heart rate monitor, pulse oximeter, or feeling ones radial wrist pulse. Heart rate tells us to some degree how hard we are exercising and how hard our heart is working. Heart rate is determined by a complex interaction with our nervous system that involves chemoreceptors (oxygen and carbon dioxide), baroreceptors (blood pressure), and the conduction system in the heart. The heart is connected to the brain, but subconsciously. An isolated heart muscle fiber will contract rhythmically on its own, but for the entire heart to pump efficiently a neural conduction system coordinates and times the contractions. Like other nerves in the body, the conduction system in the heart is very reliant on mitochondrial energy. Because the heart responds to exercise demands by changing its rate of contraction and because heart rate is easy to measure, there are many training programs based on exercising within certain heart rate ranges (zones). Ejection fraction (pump efficiency) and stroke volume (blood moved per pump) are harder to measure and generally involve high tech medical imaging to show how efficiently the heart is contracting. Ejection fraction and stroke volume are more important than heart rate when gauging the health and fitness of the heart because they measure heart muscle power. Heart muscle is similar to skeletal muscle in that it does work by contracting, is powered by mitochondria, and can get stronger with physical training. With training that is challenging enough to get the heart beating faster and stronger, the heart responds in several ways. The muscular walls of the left ventricle of the heart become thicker (hypertrophy) yet remain pliable, the left ventricular chamber becomes larger (between contractions, allowing greater filling), and the heart can empty more completely as it squeezes and contracts. Thus with each beat, the heart is moving more blood in a more efficient manner. Some elite athletes develop such efficient hearts that their resting heart rates drop to levels that would be considered abnormally low in untrained people (below 50 beats/min). Along with this increase in heart muscle strength and efficiency, heart muscle mitochondria also increase in number and efficiency. Heart mitochondria rely mainly on fat as fuel, but can also use carbohydrates and ketones. The heart must be fuel flexible, because it can never stop its vital work. Ischemic heart disease from arteriosclerosis is the leading cause of death worldwide. An ischemic heart will become weak, have fewer mitochondria, have thinner walls with scarred and stiff areas, and dilate with a big yet boggy chamber than cannot empty efficiently. In the case of high blood pressure or aortic valve stenosis, the heart is working hard all the time and may hypertrophy abnormally, with an overly thick wall and a small inefficient chamber, which is also unhealthy. Before starting an exercise program, check with your doctor to make sure your heart is healthy enough to exercise. For most of us, exercise the best thing we can do to prevent heart disease. However, there are some dangers in engaging in strenuous exercise if your heart is already diseased. Your doctor can perform certain tests (such as an exercise stress test) to see if your heart is capable of exercise.

The lungs are another key part of oxygen delivery. As we exercise, we breathe faster and more deeply. To some degree we can increase our breathing efficiency by using and strengthening our diaphragms and rib cage muscles. But unlike our muscles or heart we cannot grow new lung tissue. Oxygen and carbon dioxide exchange takes place in delicate tiny air sacs called alveoli. The total surface area of our alveoli has been likened to the surface of a tennis court. The more alveoli we have the more efficient we are at taking in oxygen. Some gifted athletes are born with larger lung capacities. Five time Tour de France winner Miguel "Big Mig" Indurain was taller than typical cyclists, but he also had a large lung capacity of 7.8 L (average is 6 L) as well as a very high cardiac output. Since you cannot grow new alveoli, you must protect the lungs you have. It is important to avoid diseases, air pollution, and infections that can damage and scar your lungs. Cigarette smoking is very damaging to the lungs and in emphysema causes irreversible death of alveoli. In emphysema, the tiny delicate air sacs are replaced with big cavities that are dead spaces that do not exchange gas. The death of alveoli in emphysema is thought to be mediated by mitochondria in a process called apoptosis.

The red blood cell (RBC) is a major component of oxygen delivery. The hemoglobin molecules packed in our red blood cells transport the oxygen to our muscles and organs and remove carbon dioxide. The redness of arterial blood comes from oxygen binding the iron in hemoglobin. This redness can be measured non-invasively using light with a pulse oximeter clamped on one's finger. The degree of redness tells us the oxygen saturation in your blood stream. In general, the more red blood cells and more hemoglobin you have, the more oxygen you can deliver to your muscles and the better you can perform. Red blood cells come from the stem cells in your bone marrow and begin with nuclei and mitochondria like other cells. As red blood cells develop they fill with hemoglobin, the molecule that carries oxygen. Hemoglobin synthesis and RBC production is a complicated multi-step process that requires many essential nutrients including: iron, vitamins B_5, B_6, B_9, B_{12}, vitamin C, vitamin E, magnesium, and zinc. When it comes to dietary support of red blood cells, most athletes only consider iron, but supporting RBC production is more complicated. Anemia is when there is not enough hemoglobin and RBC's in the blood. This can be measured by sampling the blood for hemoglobin and hematocrit (RBC percentage of blood). The good news is that a daily multivitamin will support this process. Hemoglobin is made of an iron-containing component (heme) and a protein component (globin). The heme component is synthesized entirely inside of mitochondria and is yet another important reason why mitochondria benefit health and athletic performance. Elemental lead (Pb) interferes with a key step in heme synthesis and can cause anemia. The full effects of lead poisoning on athletes will be detailed in Chapter 8: When Things Go Wrong. In sickle cell disease, there is a genetic defect in hemoglobin that causes red blood cells to deform and clog capillaries. People with sickle cell trait (having just one of two defective genes) were better able to survive malaria (rupture of the RBC prevents the malaria parasite from reproducing) and is the reason why the defective gene arose in malaria-endemic areas of the world. In 1949, Linus Pauling discovered that sickle cell disease was a genetic mutation resulting in a specific abnormal protein, making it the first disease described at the genetic and molecular levels. This was a landmark application of molecular biology in the medicine of genetic disease [31].

The maturing red blood cell sheds its nucleus that allows more room for hemoglobin and allows the RBC to have a biconcave disc shape. This semi-donut shape is maintained by certain structural proteins and allows the red blood cell to deform and fold as it squeezes through tight capillaries and this shape also increases surface area for rapid gas exchange. RBC's only have about 0.2 s to unload their oxygen inside capillaries and the deformation of the RBC and a high-range hematocrit (45–50%) facilitate this process. Mature red blood cells also lose their mitochondria (and with it their aerobic metabolism), so ironically the cells that have the most oxygen in the body do not use oxygen themselves and instead use anaerobic glycolysis. An analogy would be the diesel tanker trunk that delivers liquid oxygen to NASA, without using its own precious cargo. Red blood cell production (erythropoiesis) is increased by a kidney hormone called erythropoietin (EPO). Living at high altitude where the oxygen is thinner causes physiologic adaptations including increase in natural EPO and an elevated hematocrit. Some patients with renal failure can have weakness from anemia and may benefit with injections of EPO. Because of the effects of EPO on oxygen delivery, some athletes have cheated with illegal injections of EPO to artificially boost their hematocrit. EPO has also been found to trigger mitochondrial biogenesis [32], making the heart and skeletal muscle artificially more powerful, sometimes to super human levels. Not only is this use illegal and unethical, a hematocrit above the normal range runs the risk of blood clots, heart attack, and stroke.

Arteries are the plumbing that carries oxygenated blood to your muscles and organs, but unlike the pipes in a house, your arteries are responsive and dynamic. Arteries can shunt more blood to where it is needed the most such as working muscles; they dilate in response to exercise (lowering resistance and increasing flow); and they assist forward movement of the blood by pulsating with muscular walls. Even though the heart pumps more forcefully during exercise, the arteries dilate and reduce resistance, thus lowering blood pressure. Some patients with high blood pressure can lower their resting blood pressure with regular exercise. The arteries are lined with a delicate and metabolically active lining (the endothelium) which when healthy is thin, pliable, and smooth, but when diseased becomes thickened, hardened, and covered with plaques. The first arteries from the heart (the coronary arteries) supply blood to the heart and are especially prone to plaque formation (atherosclerosis). The arterial narrowing and blockages that result from atherosclerosis can cause weak heart muscle, disability, heart attacks, and death. Atherosclerosis in the extremities can cause peripheral vascular disease, which can result in disabling leg cramps with walking (claudication) that can progress to severe pain at rest, gangrene, and amputations. Exercise, a healthy diet, smoking cessation, and cholesterol control can help prevent atherosclerosis, improve quality of life, and increase longevity. The veins return blood to the heart and refill the chambers assisting in cardiac output. Veins do not have muscular walls and instead rely on skeletal muscle contraction and one-way valves to move blood. Inactivity such as prolonged bed rest and long distance driving or air travel can lead to venous clots (deep vein thrombosis, or DVT) that can cause limb swelling and dangerous blockage of lung arteries (pulmonary emboli). Staying active and maintaining good muscle tone prevents these clots from forming.

Skeletal muscle

There is more muscle in our bodies than any other type of tissue (except in cases of morbid obesity when there might be more adipose tissue), and all muscles use mitochondrial ATP energy. There are different types of muscle in the body: smooth muscle (involuntary) found in the lining of the gastrointestinal tract and arteries, striated cardiac muscle (involuntary) as mentioned previously, and striated skeletal muscle (voluntary). In this section, we will discuss the latter, the voluntary skeletal muscles you control during exercise and sports. Skeletal muscles are striated because when seen under the microscope they have overlapping transverse bands that change in width during contraction. The bands are composed of actin and myosin. Actin is thin and rope-like, and myosin has grabbing and flexing arms. When signaled to contract by the nervous system, calcium is released and the myosin arms are powered by ATP to grab, pull, release, grab, pull, release, grab, and pull the actin ropes. Within each muscle there are trillions of myosin filaments that use quadrillions of ATP when exercising. Mitochondria convert fuel (carbohydrates, fat, and amino acids) into the useful ATP energy that powers our muscles. To account for the different types muscular effort (e.g., strength vs. endurance) there are different types of muscle fibers, notably different in their mitochondrial density. Slow twitch (Type I) muscle fibers engage during endurance activities, have a high mitochondrial density, are darker in color from the heme in mitochondria and myoglobin, require oxygen (myoglobin temporarily stores limited oxygen), and are resistant to fatigue. Fast twitch (Type II) muscle fibers engage during short duration strength activities, have a lower mitochondrial density, are lighter in color (less heme), can work with or without oxygen, and fatigue easily.

An example of different muscle fiber types is the white meat and dark meat in chickens. The dark thigh meat in chickens is composed of endurance slow twitch muscles rich in heme and mitochondria, because chickens spend their waking hours walking as they peck at food on the ground. The white breast meat in chickens is powerful fast twitch muscle with fewer mitochondria, because chickens only need to fly in short bursts to evade predators. The breast meat in migratory birds such as ducks and geese is composed mainly of dark muscle meat (slow twitch and mitochondria-rich) because these birds must fly for hours at a time. Human muscle is more of a mixture of these different fibers and in cross section looks like a mosaic. Some muscles such as the biceps in the arms are mainly fast twitch fibers, some muscles such as the postural back muscles are mainly slow twitch fibers, and other muscles such as the calf muscles are a combination of both types of fibers. There are also intermediate muscle fibers that with specific training can be recruited to be more fast twitch or more slow twitch. There is some debate as to whether children are born with a muscle fiber distribution that locks in specific sports potential (such as the position on a football team). There is no debate that by training to increase your mitochondrial mass and to the better supply oxygen to your mitochondria, the better you will perform. Muscle performance will be further detailed in Chapter 6: Your Body at Work, Play, and Rest, Chapter 10: Gauging Fitness, and Chapter 11: BEAST Fitness Training.

What it takes to win the Tour de France

The Tour de France is the most challenging and grueling athletic event in the world. It has been compared to racing a marathon every day for 3 weeks. Contested over 2000 miles (3200 km), in 21 days of racing, against some of the fittest competitors in the world, the "Tour" requires mastery of several disciplines. The winner must have the power-to-weight ratio to climb mountain passes, the steady power to win time trials, the strength to win final sprints, and the endurance to vie at the end. The Tour selects an all-around cyclist, not a specialist. Greg LeMond was a true all-around athlete who could climb with the best, beat everyone in the time trials, won world championships with his finish line sprint, and seemed to grow stronger at the end of long stage races. In his first Tour de France he finished 3rd, in his second Tour he finished 2nd, and in his third Tour he won. LeMond had the right combination of muscle power to be the best at many disciplines. We contend that during his era, LeMond also had the best mitochondrial energy to power those muscles.

Homeostasis: fluid and temperature balance

The body operates most efficiently within very narrow parameters, and maintaining conditions within these parameters is called homeostasis. Keeping things steady should not be confused with keeping things equal. For example, body temperature homeostasis means keeping the human body steady at 98.6°F (37.0°C) and does not mean keeping the body equal to room temperature. If our body temperature is either just a few degrees higher (fever) or a few degrees lower (hypothermia) than normal, we feel very miserable. By far the most important homeostatic condition is something most athletes know very little about, the sodium (Na^+) and potassium (K^+) salt gradient across our cell membranes. Imbedded in every cell membrane in the body are thousands of Na^+/K^+ pumps, which maintain this homeostasis [33]. These tiny pumps are channels that move Na^+ ions out of the cell (three at a time) and K^+ ions inside the cell (two at a time). Moving these ions from low concentration to high concentration takes energy in the form of one high energy ATP being converted into ADP. These tiny pumps are so numerous that during rest they are the biggest consumer of ATP in the body. Much of our mitochondrial energy generation is dedicated to powering these tiny pumps. This state of salt imbalance is important for many reasons. It keeps the cell properly hydrated because water molecules surround each Na^+ and K^+ ion. Without the Na^+/K^+ pumps, the cell would swell and burst. The imbalance also drives the movement of critical nutrients (such as glucose and amino acids) across cell membranes by means of other imbedded transport membrane proteins that are powered by the Na^+/K^+ imbalance. The imbalance also regulates the flow of calcium (Ca^{++}) in and out of cells for nerve transmission and muscle contraction. Just as these tiny pumps maintain salt and hydration homeostasis inside the cell, our kidneys maintain salt and hydration homeostasis outside the cell. The microscopic working units of our kidneys are called nephrons and keep our body fluids in control. The kidneys filter and reabsorb our water and eliminate excess salt. They also eliminate excess metabolites such as urea, lactic acid, excess protein, and excess glucose. When blood glucose is high, the kidney can eliminate some of it into the urine. Every glucose molecule comes with six molecules of water, so diabetics have sweet and copious urine. This loss of water makes diabetics very thirsty. Symptoms of diabetes include polyuria (excessive urination) and polydipsia (excessive thirst). Other key parts of hydration homeostasis include the digestive tract and sweat glands.

When we exercise some of the energy creates mechanical force (muscle power) but some of the energy creates heat. To maintain body temperature during exercise, it is important to shed this heat. When we sweat, we lose water and salt through our sweat glands and as the water evaporates, the hottest water molecules leave the skin surface and take away heat. There is a reason why sports drinks contain sugar and salt. The sugar speeds hydration because six molecules of water surround each molecule of glucose, and the salt replaces what is lost in sweat. Cooling can be facilitated with adequate hydration, clothing that wicks water away, and air movement or breezes that increase evaporation. When heat cannot be lost, such as when a small child or pet is trapped inside a car in the summer sun without water or breeze, the body temperature can rise to dangerous levels. Heat cramps occur when the temperature in muscle is high enough to impede muscle contraction. This can be followed by symptoms of heat exhaustion: profuse sweating, dizziness, weakness, pale skin, lack of urination, confusion, and a rapid resting heart rate. If allowed to progress further, this can lead to heat stroke, a condition when the body's cooling systems fail with symptoms including: failure to sweat despite heat, red dry skin, throbbing headache, severe weakness, fainting, seizures, and unconsciousness. If heat stroke is not treated immediately by cooling the body, the kidneys, muscles, and brain can be damaged, leading to coma and death.

A more insidious temperature problem with athletes is hypothermia, i.e., being too cold. If athletes do not prepare for cold weather or there is an unexpected drop in outdoor temperature, they can be at risk for hypothermia. In the Arctic Circle, savvy outdoors people are careful not to overexert themselves when away from shelter at dusk. If they become too sweaty with intense work as the temperature falls, evaporation can cause a rapid drop in body temperature. The symptoms of hypothermia include: paleness (skin vasoconstriction), shivering (muscle contraction generates some heat), and confusion (that can lead to poor survival decisions). This can progress to severe shivering, cyanosis (blue lips), loss of coordination, and irrational thinking. If not treated immediately by warming the body, the heart rate, breathing, and blood pressure can all decrease as the body shuts down. Symptoms of severe hypothermia include stumbling, loss of motor control, difficulty speaking, organ failure, and death. As noted in the previous section on body fat, there is a special type of fat called brown adipose tissue that can generate pure heat from body fat. This is done without shivering or muscle contraction. Studies have shown that this body warming and fat burning effect is triggered by exposing the extremities to cold. Although some have tried ice baths to encourage their brown adipose tissue, this practice is impractical, dangerous, painful, and unnecessary. A brisk walk in cold weather or a brief cold shower may be all you need to activate your brown adipose tissue.

The immune system

Our bodies are in a constant state of war in a battle against invasion. To stay healthy and to perform optimally we need to be on the winning side of this battle at all times. Every surface we touch and the people we meet are all covered with microbes (bacteria, viruses, and other germs). Studies have shown that a computer keyboard has more germs than a public toilet seat [34]. Bacteria cover every inch of our skin and are always trying to enter our bodies.

Hand washing

One of the first experiments we had in medical school was to imprint our hands on a Petri dish before and after washing our hands. My lab partner had just come from the bathroom and said that his freshly washed hands would ruin the experiment. I suggested that he run his hands through his hair (visible clean) one time to "soil" his hands. The next day that "soiled" imprint left a sharply demarcated hand print of bacteria. The post hand wash Petri dish was almost completely clean. This experiment showed that hand washing is very important in controlling the spread of disease.

Mark Hom

The dry layer of dead skin cells on our skin serves as a barrier to infection. Germs thrive in moist conditions and the moist lining of our nostrils is a perfect environment for bacteria. A common test in modern hospitals is to swab patients' nostrils to check for bacterial colonization. Hospital acquired infections often are due to antibiotic resistant bacteria that are difficult and expensive to treat. Current health care laws penalize hospitals for causing these infections, so the swab is often used as proof that the patient caused his own infection. Because hospital staff are exposed to these resistant germs on a daily basis, they are often the carriers and therefore hand washing is the best thing medical professionals can do to control hospital infections.

Other barriers to infection include mucus (to clear germs from the nose, mouth, and lungs), gastric acid that kills many of the germs that might get into our food, and the lymph nodes along our digestive tract. Having bacteria in your colon is normal, but it is important to have healthy bacteria to keep harmful bacteria in check. An analogy is having a healthy lawn of grass to suppress weed growth. Until recently it was a mystery as to why we have the appendix, a short blind ending tube of intestine at the start of our colon. When inflamed it can cause appendicitis, a medical emergency requiring surgery. Doctors wondered why we have them at all. The latest theory is that the appendix carries a reserve of good bacteria and if the colon is overwhelmed by bad bacteria, the appendix can jump start the colon with good bacteria [35]. This is the theory behind probiotic foods that encourage healthy colon bacteria. It is also the theory of stool transplant, a disgusting yet effective way to treat antibiotic resistant forms of chronic bacterial diarrhea [36].

The oral cavity is another source of bacteria and the teeth and gums must be taken care of or else infection can travel from the mouth to the blood stream. Good dental care correlates with less chronic inflammation and better health.

When infection penetrates our barriers, the white cell component in our blood attacks the invaders with antibodies, dissolving enzymes, and other means of chemical warfare. Sometimes the immune system gets confused and attacks the body it is meant to protect. Autoimmune disease include: type 1 diabetes, celiac disease, Crohn's disease, ulcerative colitis, rheumatoid arthritis, lupus, eczema, psoriasis, vasculitis, Wegener's granulomatosis, and dozens more.

The most insidious infection of all is from acquired immunodeficiency syndrome, human immunodeficiency virus (AIDS/HIV) which attacks and weakens the immune system itself, making it difficult to clear the body of the viral infection, but also making the body vulnerable to many viral, bacterial, and parasitic infections that would normally not cause great problems.

All forms of chronic inflammation whether from bacteria, viruses, and autoimmune disorders can weaken the body, siphon away vital energy, and accelerate aging and degeneration in the body. This is why athletes need to prevent infection and maintain a strong immune system.

The brain and nervous system

The last organ system in athletic performance is the brain and nervous system. By commanding the muscles to act, the brain controls every aspect of the athlete's speed, power, and endurance. By sensing with vision, hearing, touch, and pain, the brain reacts and adjusts to the situation. The brain also strategizes, communicates, and follows the game plan. It was once believed that the brain was a cold lump of fat with a fairly low metabolic requirement. We now know that to perform all of these neurologic functions, the brain uses large amounts of mitochondrial ATP energy. On a modern PET scan of a person at rest, the two organs with the highest metabolic activity are the brain and the heart. ATP powers the transmission of electrical nerve impulses and powers calcium gradients in the release of neurotransmitters at the nerve ends (synapses). Despite its great power, the brain is a soft and delicate structure that is protected by the skull and a bath of liquid (cerebrospinal fluid). Between the capillaries and the brain there is a physical cellular wall called the blood brain barrier that protects the brain from foreign substances, bacteria, and large toxic chemicals. Because of the blood brain barrier, only small fuel molecules can cross (many by active transport), such as sugar (and ketones during starvation). Long chain fatty acids are too large cross the blood brain barrier and although the brain contains structural fat to insulate its wire-like connections, it does not store fat for energy. Because of these unusual features, the brain has a very unique energy metabolism based on carbohydrates (mainly sugar and a limited supply of glycogen). Often called the selfish organ, the brain makes sure its energy needs are met before all other organs [37]. Whereas the brain's sugar craving contributes to obesity in sedentary people, an athlete can eat lots of carbohydrates yet burn them off, keeping the mind and body in better harmony. It was once a mystery as to why we require sleep but the current theory explains sleep as a reenergizing process of the brain's energy storage system. When we are sleep deprived we cannot think clearly, we cannot concentrate, we make bad choices, and we make mistakes. It was known previously that the brain contained glycogen, but because the quantities are so small (compared to muscles and the liver), brain glycogen was thought to be insignificant. During waking hours the brain uses up this limited supply of stored glycogen. As this occurs adenosine accumulates and creates a gradual feeling of drowsiness and makes us want to sleep. It was found that when we sleep the brain re-accumulates its glycogen (just as a cell phone will recharge faster when turned off) [38]. This re-energizing process explains why we require sleep, why we feel refreshed after a good night's sleep, and why athletes perform better when not sleep deprived. The active ingredient in coffee (caffeine) blocks adenosine receptors and staves off the drowsy feeling. Caffeine makes us more alert and also has a mild ergogenic effect on muscles. However, coffee interferes with normal sleep and the brain's glycogen rebuilding process. Although coffee in the morning can be a minor help in mental and physical performance, it should be avoided by athletes in the evening and at night. Another example of the selfish brain is during starvation, when our muscles and liver lose their glycogen, while the brain protects its glycogen stores. During prolonged starvation the brain relies on ketones, produced by the liver from fat.

The brain is the master control organ and is hard wired to the other organs via the peripheral nervous system. Skeletal muscle is controlled voluntarily from the motor cortex of the brain. Much of the brain's control of muscle is inhibitory, i.e., control rather than jerky motion. A good athlete uses conservation of effort and does not waste motion or energy. As this signal travels down the brain stem, it is modulated and coordinated by the cerebellum which gives the athlete a sense of balance and agility. Many other systems are controlled autonomically (without conscious thought) by the autonomic nervous system. Our heart, arteries, intestines, irises, and skin pores are controlled by the nervous system but without conscious thought. Some processes such as breathing have both voluntary control (breath holding) and autonomic control (breathing while you sleep).

The brain needs a constant supply of oxygen to power its many mitochondria. Interruptions in the oxygen delivery can result in permanent ischemic injury to the delicate brain tissue. Because the brain is so energy dependent, mitochondrial energy has a big role in brain function and faulty mitochondria are thought to be the center of many degenerative brain diseases. We will discuss brain energy further in the Chapter 7: The Body–Brain Connection. We will discuss mitochondrial implications in brain disease in Chapter 8: When Things Go Wrong and in Chapter 9: Slowing the Aging Process (Figure 2.4).

Figure 2.4 "Life is like riding a bicycle. To keep your balance you must keep moving."
Albert Einstein.
Image by Mark Hom.

Why biology?

When most people try to remember what they learned in high school biology, they recall the smell of frogs in formaldehyde, memorization of strange words like "endoplasmic reticulum," and endless categorization and subcategorization. While it is true that much of biology is dissecting things down to fundamental components, the underlying beauty of biology is its unifying principles such as genetics, evolution, and the interconnectedness of life. Fitness at its core is a biologic process and mitochondrial biology is the unifying principle behind fitness. Many fitness books are pseudoscientific, borrowing terms from biology but skirting around the true basis of physical performance. We are told to exercise with intensity but we are not told why that works. We are told that fitness is a building process, but we are not told what exactly is being built. We are told that physical training develops power, but we are not told the source of that power. The biology of mitochondria answers these questions and many more.

Why does everyone need to know biology? Simply, biology is the owner's manual for the human body. If you are aware of biology, you know the correct fuels to put in the tank, how to attend to your body's needs, what to expect as the miles accumulate, and how to make your body run better and longer. "The Science of Fitness" is the athlete's owner's manual. If you want to understand your body and how to make it perform at its best, we encourage you to read your owner's manual cover to cover (Figure 2.5).

Figure 2.5 When most people remember high school biology they think about dead frogs in formaldehyde, but biology is the study of life.
Photography: Kirill Tsukanov.

Inside the cell

Exercise, at its core, is a biological process. Up to this point we explained athletic performance as a physiological collaboration of various organ systems, each with a special role in energy metabolism and athletic power. Now we will delve into yet a deeper biological level – inside the cell itself. The athlete who can appreciate and apply what is known about cell biology will have a great advantage in achieving his or her fitness and endurance goals. Currently, the vast majority of athletes train without really thinking about what really goes on inside their cells. Sure, athletes think about contracting muscles, since muscles are the focus of most fitness books and the obvious end product of performance. Aching muscles are what the athlete feels during intense activity and during recovery. However, muscles are composed of cells (in this case, cells specialized in converting cell energy into physical motion). What's more, other types of cells produce, store, and regulate the chemical energy that drives muscle contraction.

Here, we will reveal what is under the veil of the cell, focusing on the major components of the cell that greatly impact athletic performance. By learning these few concepts from cell biology you will soon grasp topics such as genetic athletic gift and individuality, genetic expression through protein synthesis, carbohydrate uptake, fat burning, vital cell energy functions, and the power that drives performance. Although the cell is dissected here for exploratory purposes, you will also soon appreciate that the living cell – the fundamental unit of life – is truly greater than the sum of its parts. As we describe the various functions of the cell, the reader may get the idea that each cell is like a one man band with 12 instruments strapped to his sides in a comical fashion. A better analogy would be to think of the cell as a full orchestra, conducted by the nucleus and with specialized organelles performing specific roles, yet all harmoniously playing the same song of life (Figure 2.6).

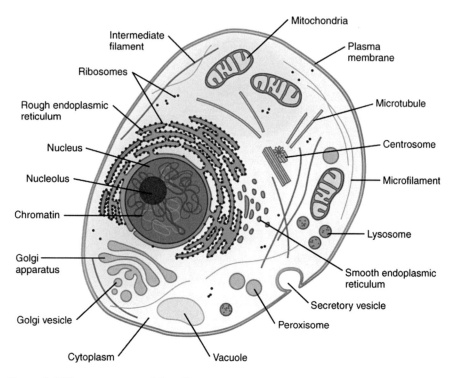

Figure 2.6 The components of the cell.
Image by OpenStax College.

The gate keeper

The cell membrane is the outer casing that regulates what enters and exits the cell. The membrane is "selectively permeable," a term that means it allows passage of some molecules but not others. For the most important molecules of metabolism, such as glucose and ATP, there are enzymes imbedded in the cell membrane that actively transport these substances. Nutrient transport, hydration, cell communication, and cell integrity are functions of the cell membrane. Indeed, the membrane is a dynamic and responsive entity, not merely a concrete wall.

Membranes are composed of phospholipids (fat) and imbedded proteins and thus (when being constructed) require healthy dietary fats and protein. Although most of us have a negative association with fat, for good reason, fat is also an important component of our bodies and vital in the functioning of cell membranes. Indeed, some dietary fats, such as certain fish oils and olive oil, are beneficial to cell membrane functioning. On the other hand, some unhealthy dietary choices such as artificial trans fats (trans fatty acids) can distort and weaken cell membranes. In Chapter 8: When Things Go Wrong, we will explain the nuances of reading a nutritional label so that athletes can eliminate trans fats from their diets.

Our blueprint

In the center of the cell lies the nucleus – the home of our DNA. DNA is the blueprint from which all of our proteins, and ultimately all of the structures in the body, are created. In its entirety, our nuclear DNA is referred to as our genome. The DNA code in our genome is a combined product of both of our parents and is what determines much of our individuality: from our height and eye color to our various strengths and weaknesses. One vital function of the nucleus is to protect the genome from mutation by using protective packaging and repair mechanisms. The nucleus also controls and coordinates important cell functions such as cell division, and is often said to be the manager that, "keeps a lid on things." Cancer is one example where the central regulation fails, critical cell functions are impaired, and cell division goes out of control.

Our genome plays a major role in determining our athletic abilities. It is our "genetic gift" presented to us at conception by our parents. Although it is easy for athletes to blame genetics for their limitations, our genetic gift is also the source of our athletic potential (largely under-tapped in most people). To make the most of our genome, we need to affect "genetic expression" – the conversion of DNA code into performance. An athlete's genetic expression is impacted by training, proper diet, and a variety of environmental factors. As will be detailed in the next chapter, mitochondria have their own separate DNA, which is inherited in a different pattern compared to nuclear DNA. Therefore, mitochondria also contribute to an athlete's genetic gift.

The protein factory

Much of our body (including hair, collagen, enzymes, and muscle fibers) is composed of proteins which in turn were created from the instructions encoded in our DNA. Within the cell lies an important "protein factory" – the endoplasmic reticulum – where DNA instructions are followed and proteins are carefully assembled. Proteins are complicated structures, formed by chains of various amino acid building blocks. Although our bodies can synthesize many of the amino acids that make up proteins, some amino acids are "essential" – that is, we cannot synthesize them, so they must be included in our diet. Without the intake of essential amino acids, the endoplasmic reticulum cannot properly produce all the proteins we need. For an athlete this need is amplified, because athletes rebuild their muscles and cells at a rate higher than non-athletes. At one time, athletes were obsessed with carbohydrate loading to fuel muscles. Indeed this was once the main tenet of sports nutrition. We now know that a more balanced approach will better supply what the cells demand, not just for energy, but for other vital processes as well. The current thinking in sports nutrition incorporates carbohydrates, protein, good fat, electrolytes, vitamins, minerals, and micronutrients in order to feed the cell and supply the cell's protein factory.

The mailroom

After the endoplasmic reticulum manufactures our proteins, a structure called the Golgi apparatus takes the freshly synthesized proteins and then folds, packages, and labels them for transport inside and outside the cell. One of the more recent discoveries about proteins is that they are much more than a mere sequence of amino acids; they are complex three-dimensional superstructures. Only by folding into the correct shape will a protein become fully functional. This folding can be so intricate that it requires a network of super computers to simulate. Mad cow disease is an example of protein folding gone wrong. The infectious agent in mad cow disease is neither a bacterium nor a virus, but a prion – a misfolded protein that causes other proteins to fold incorrectly. We rely on the Golgi apparatus to perform its complicated functions, so our proteins head off in the right direction and complete the tasks for which they were created. We do not currently understand all of the proteins or even how many proteins our bodies make for the various human functions, but proteins are our genetic expression.

The recycling centers

Cells are constantly rebuilding their inner machinery of proteins and lipids. Just as an old or well-used computer accumulates corrupt files over time, a cell accumulates worn out and corrupt molecules that must be recycled and replaced. An example is when the high glucose levels in diabetes result in glucose spontaneously binding to proteins, making them dysfunctional. This, combined with free radical damage, explains part of the vicious cycle in diabetes where high sugar levels result in poor metabolism and organ damage. Two types of organelles in the cell, lysosomes and peroxisomes, recycle worn out proteins by collecting them and dissolving them into their building blocks for the synthesis of new proteins. Lysosomes use enzymes and acid to digest worn out cell components, from single molecules to entire organelles. Peroxisomes use hydrogen peroxide to detoxify the cell by ridding it of harmful chemicals. Peroxisomes also have a role in lipid metabolism and prepare fatty acids for fat burning in mitochondria.

The power plants

Lastly, but arguably most importantly for an athlete, are the intracellular dynamos called mitochondria (pleural for mitochondrion). Through a complex symphony of enzymatic reactions, mitochondria convert carbohydrates, fats, and proteins into ATP. ATP is the chemical energy currency that the cell uses to power the body, from chemical reactions to muscle contraction. Any sort of sustained athletic effort requires mitochondrial energy. Mitochondria can convert body fat or digested food into the energy that powers muscles. In the process of producing energy, mitochondria consume oxygen. Indeed, the best laboratory test of an athlete's fitness is measurement of oxygen consumption – the burning that occurs in our mitochondrial fire. These chemical reactions generate harmful free radicals, which must be neutralized by mitochondrial antioxidant systems. The beneficial antioxidants found in fruit and vegetables act at the cellular and mitochondrial level.

Mitochondria have the amazing ability to multiply in order to meet the cell's energy demands. If we were to look inside the cells, we would find many more mitochondria in an elite athlete than in a recreational athlete, who in turn would possess many more mitochondria than a sedentary person. Superior athletes were not necessarily born with more mitochondria; more are produced in response to training, and

Figure 2.7 In a record setting time trial performance of 33.9 miles/h (54.5 km/h) Greg LeMond won the 1989 Tour de France by a mere 8 s margin on the final day. Photography: AFP/Getty Images.

the more vigorous the training, the more mitochondria are produced within the cell. Mitochondrial multiplication (biogenesis) is triggered by the energy sapped state that results from intense physical training. It turns out that mitochondria multiply in a very economical fashion, only increasing in number when more power is needed. One major goal of this book is to explain why high-intensity interval training is crucial in maximizing mitochondrial energy production. While other fitness books may recommend some of the training methods we advocate, our SciFit approach explains training at the biological level so you will understand *why* certain training methods work better. The right type of exercise will trigger mitochondrial biogenesis – thus increasing the number and energy efficiency of your mitochondria, and making you a better athlete (Figure 2.7).

References

[1]   Wada H, Satoh N. Phylogenetic relationships among extant classes of echinoderms, as inferred from sequences of 18 S rDNA, coincide with relationships deduced from the fossil record. Journal of Molecular Evolution 1994;38:41–9.

[2]   Powell H. Dinosaur vs. Crocodile: Who Wins? www.smithsonianmag.com; 2008.

[3]   McDougall I, Brown FH, Fleagle JG. Stratigraphic placement and age of modern humans from Kibish, Ethiopia. Nature 2005;433(7027):733–6.

[4]   Perko MJ, Nielsen HB, Skak C, et al. Mesenteric, coeliac and splanchnic blood flow in humans during exercise. Journal of Physiology 1998;513(3):907–13.

[5]   van Nieuwenhoven MA, Vriens BE, Brummer RJ, Brouns F. Effect of dehydration on gastrointestinal function at rest and during exercise in humans. European Journal of Applied Physiology 2000;83(6):578–84.

[6]   Ma T, Verkman AS. Aquaporin water channels in gastrointestinal physiology. Journal of Physiology 1999;517:317–26.

[7]   Carpenter GH. The secretion, components, and properties of saliva. Annual Review of Food Science and Technology 2013;4:267–76.

[8]   Davis JM. Fluid availability of sports drinks differing in carbohydrate type and concentration. American Journal of Clinical Nutrition 1990;51(6):1054–7.

[9]   White JS. Straight talk about high-fructose corn syrup: what it is and what it ain't. American Journal of Clinical Nutrition 2008;88(6):1716S–21S.

[10] Larsen ML, Hørder M, Mogensen EF. Effect of long-term monitoring of glycosylated haemoglobin levels in insulin-dependent diabetes mellitus. New England Journal of Medicine 1990;323(15):1021–5.

[11] Williams PG. Nutritional composition of red meat. Nutrition & Dietetics 2007; 64(Supplement 4):S113–9.

[12] Lucan SC. That it's red? Or what it was fed/how it was bred? The risk of meat. American Journal of Clinical Nutrition 2012;96(2):446.

[13] Uzel C, Conrad ME. Absorption of heme iron. Seminars in Hematology 1998;35(1):27–34.

[14] Kris-Etherton PM, Harris WS, Appel LJ. Fish consumption, fish oil, omega-3 fatty acids, and cardiovascular disease, AHA Scientific Statement for the Nutrition Committee. Circulation 2002;106:2747–57.

[15] López-Miranda J, Pérez-Jiménez F, Ros E, et al. Olive oil and health: summary of the II international conference on olive oil and health consensus report. Nutrition, Metabolism and Cardiovascular Diseases 2010;20(4):284–94.

[16] Kozimor A, Chang H, Cooper JA. Effects of dietary fatty acid composition from a high fat meal on satiety. Appetite 2013;69:39–45.

[17] Anderson JW, Baird P, Davis RH Jr, et al. Health benefits of dietary fiber. Nutrition Reviews 2009;67(4):188–205.

[18] Zeng H, Lazarova DL, Bordonaro M. Mechanisms linking dietary fiber, gut microbiota and colon cancer prevention. World Journal of Gastrointestinal Oncology 2014;6(2):41–51.

[19] Sanyala AJ, Yoonb SK, Lencionic R. The etiology of hepatocellular carcinoma and consequences for treatment. The Oncologist 2010;15(Suppl. 4):14–22.

[20] Djoussé L, Gaziano JM. Dietary cholesterol and coronary artery disease: a systematic review. Current Atherosclerosis Reports 2009;11(6):418–22.

[21] Olson RE. Discovery of the lipoproteins, their role in fat transport and their significance as risk factors. Journal of Nutrition 1998;128(2):439S–43S.

[22] Theuwissen E, Mensink RP. Water-soluble dietary fibers and cardiovascular disease. Physiology & Behavior 2008;94(2):285–92.

[23] Katz J, Tayek JA. Gluconeogenesis and the Cori cycle in 12-, 20-, and 40-h-fasted humans. American Journal of Physiology 1998;275(3 Pt 1):E537–42.

[24] Hasselbalch SG, Knudsen GM, Jakobsen J, et al. Blood-brain barrier permeability of glucose and ketone bodies during short-term starvation in humans. American Journal of Physiology 1995;268(6 Pt 1):E1161–6.

[25] Bacha TE, Luz MRMP, Da Poian AT. Dynamic adaptation of nutrient utilization in humans. Nature Education 2010;3(9):8.

[26] Frankenfield DC, Rowe WA, Cooney RN, et al. Limits of body mass index to detect obesity and predict body composition. Nutrition 2001;17(1):26–30.

[27] Jensen MD. Role of body fat distribution and the metabolic complications of obesity. Journal of Clinical Endocrinology and Metabolism 2008;93(11 Suppl. 1):S57–63.

[28] Gaesser GA. Thinness and weight loss: beneficial or detrimental to longevity? Medicine and Science in Sports and Exercise 1999;31(8):1118–28.

[29] Clarke JR, Brglevska S, Lau EW, Ramdave S, Hicks RJ. Atypical brown fat distribution in young males demonstrated on PET/CT. Clinical Nuclear Medicine 2007;32(9):679–82.

[30] Seale P, Lazar MA. Brown fat in humans: turning up the heat on obesity. Diabetes 2009;58(7):1482–4.

[31] Pauling L, Itano HA, Singer SJ, Wells IC. Sickle cell anemia, a molecular disease. Science 1949;110(2865):543–8.

[32] Carraway MS, Suliman HB, Jones WS, et al. Erythropoietin activates mitochondrial biogenesis and couples red cell mass to mitochondrial mass in the heart. Circulation Research 2010;106(11):1722–30.

[33] Skou JC, Esmann M. The Na, K-ATPase. Journal of Bioenergetics and Biomembranes 1992;24:249–61.

[34] Childs D. Your keyboard: dirtier than a toilet. ABC News Medical Unit; 2008.

[35] Bollinger RR, Barbas AS, Bush EL, Lin SS, Parker W. Biofilms in the large bowel suggest an apparent function of the human vermiform appendix. Journal of Theoretical Biology 2007;249(4):826–31.

[36] Bakken JS, Borody T, Brandt LJ, et al. Treating *Clostridium difficile* infection with fecal microbiota transplantation. Clinical Gastroenterology and Hepatology 2011;9(12):1044–9.

[37] Fehm HL, Kern W, Peters A. The selfish brain: competition for energy resources. Progress Brain Research 2006;153:129–40.

[38] Benington JH, Heller HC. Restoration of brain energy metabolism as the function of sleep. Progress in Neurobiology 1995;45:347–60.

Amazing Mitochondria

3

What are mitochondria?

Mitochondria inhabit the cytoplasm of our cells. The cytoplasm is the larger part of the cell, inside the cell membrane but outside of the nucleus. The name "mitochondria" comes from their appearance under a powerful microscope, from the Greek roots words mitos (thread-like) and chondros (granular). "Mitochondria" is the pleural form of mitochondrion, but we will rarely speak about a single mitochondrion in isolation as they are generally present in great numbers. These thready granules were observed to be prevalent in all of our living cells and later it was discovered that they give the cell its vitality. The main function of mitochondria is to turn the food we eat (carbohydrates, protein, and fat) into useable energy in the presence of oxygen. The energy we gain powers every thought, motion, step, heartbeat, and athletic endeavor. Without mitochondria we would literally not be able to lift a finger (Figure 3.1). The more energy we have on tap, the better our bodies can perform. This applies not only to athletic performance, but also to our vitality and health. This book focuses on mitochondria because they are the center of human energy, athleticism, and metabolism. By understanding mitochondria you can better tap your athletic potential and preserve your health. Besides generating cell energy, mitochondria also burn fat, synthesize hemoglobin, protect us from free radicals (the cause of aging), power digestion and other processes, regulate body temperature, and determine how long our cells live.

Mitochondria are masters of oxygen. In the presence of oxygen, mitochondria can convert fuel into adenosine triphosphate (ATP), the chemical energy that our cells and muscles can use. The fuels can be in the food we eat (carbohydrates, dietary fat, and meat) or in the energy reserves our body stores (glycogen, body fat, and protein). Mitochondrial biology is often taught in the class room using glucose as the prototypical fuel, but mitochondria also burn our body fat (when we exercise properly) and can also burn the amino acids from protein (in high protein diets and times of starvation). Mitochondria are like high efficiency furnaces and can convert one glucose molecule (using oxygen) into 34 ATP [1]. There is some debate as to just how many from a theoretical high of 38 ATP to only 30 ATP in suboptimal conditions, but we will use 34 as an average. Without oxygen our cells revert to anaerobic glycolysis, a far simpler process which only produces four ATP and uses two ATP for a net gain of only two ATP [1]. This is why mitochondria are so important and highly efficient. They can burn fuel in the presence of oxygen to generate 17 times more energy.

The first step in understanding how mitochondria achieve this energy advantage is to understand oxygen. A single atom of oxygen (element O, atomic number 8) has eight protons and eight electrons. Two of these electrons are unpaired, making O highly reactive because it is more stable when these electrons are shared with other atoms. O by itself is therefore very short lived and will readily combine with other

The Science of Fitness: Power, Performance, and Endurance. http://dx.doi.org/10.1016/B978-0-12-801023-5.00003-X

Figure 3.1 Mitochondria power every muscle in your body, from the largest leg muscles to the tiny muscles in the iris of your eye.
Image by Mark Hom.

nearby atoms to create other molecules such as H_2O (water), CO_2 (carbon dioxide), PO_4 (phosphate), SO_4 (sulfate), Fe_2O_3 (iron oxide, rust), O_3 (ozone), and even with itself to form O_2 (what we commonly call oxygen). O basically wants to combine with everything and this tendency makes oxygen drive many chemical reactions inside and outside the body. It was once thought that oxygen was a prerequisite for life. However, it is now believed that life on Earth originated without oxygen [2]. Oxygen would have been too reactive and resulted in far too simple chemicals, rather than the more complicated chemicals required for life to begin. The classic Miller–Urey experiment (first performed in 1953 and still performed in chemistry labs today) demonstrated that if the gases known to be present in the Earth's primitive atmosphere (water, methane, hydrogen, and ammonia, but no oxygen) are exposed to electrical spark, many organic molecules are produced including: 22 amino acids (the building blocks of proteins), nucleotides (the building blocks of DNA), and sugars [3]. Life first developed when these organic chemicals combined in shallow cesspools at the edges of the oceans. If highly reactive oxygen had been present this might not have happened. The first forms of life were anaerobic bacteria that thrived in this low oxygen environment. Some were able to use the heat energy and sulfur compounds from sea vents and others began to use solar energy. Eventually, bacteria developed that could convert water (H_2O) and carbon dioxide (CO_2) into sugar using light energy, the process of photosynthesis. Descendants of these first photosynthetic organisms exist today in the form of cyanobacteria and the chloroplasts inside plants. A byproduct of photosynthesis was O_2 which at first was captured by metals such as iron, creating rust bands in geologic formations that began about 2.4 billion years ago [4]. Once these inorganic compounds were saturated with oxygen, free atmospheric O_2 began to rise sharply. Oxygen (O_2) is currently about 20% of the air we breathe.

The introduction of oxygen into the atmosphere caused a tremendous die off of anaerobic life, often called the oxygen catastrophe [5]. Oxygen and the free radical compounds it can create were so reactive that oxygen was toxic to the many primitive forms of life. Anaerobic bacteria went from being the dominant form of life to being relegated to the mucky bottoms of ponds. On the other hand, newly evolved bacteria that could handle oxygen and use it to their advantage began to thrive. A type of aerobic bacteria related to mitochondria began to use oxygen to create more energy in a process called oxidative phosphorylation. The oxygen reactivity that was toxic to anaerobic life was now driving high-energy reactions, allowing mobility, and eventually creating the animal branch of the evolutionary tree.

Ozone: protector and pollutant

An example of the reactivity of oxygen is how oxygen, light, and ozone (O_3) interact. High above in the stratosphere, oxygen and ozone help absorb harmful light rays that would otherwise be damaging to our skin. Ordinary oxygen and nitrogen filter out ultraviolet (UV) light with wavelengths below 200 nm, whereas the ozone layer partially absorbs UV light from 200 nm to 315 nm including the damaging UV-B band [6]. Without the ozone layer our skin would burn easily and skin cancer would be a much greater threat to everybody.

$$O_2 + photon \rightarrow 2O$$

$$O + O_2 \rightarrow O_3$$

$$O_3 + O \rightarrow 2\,O_2$$

However, ground level ozone can be harmful. During hot summer months and especially in densely populated urban areas, automotive emissions of hydrocarbons and nitrogen oxides can react with sunlight and create O_3 (ozone), which is even more reactive than O_2. Ozone can then create many other highly reactive chemicals and free radicals (the dangerous components of smog) which can be irritating and damaging, especially in those with preexisting lung and heart conditions [7].

Two of the greatest leaps in evolution were photosynthesis (the capture of solar energy and production of oxygen) and oxidative phosphorylation (the use of oxygen to create more energy). Without oxygen and without mitochondria, our early ancestors would literally not have had the energy to crawl out of the primordial ooze. When you exercise, it is easy to take for granted the air you breathe, but inside every one of your cells, your mitochondria turn oxygen into athletic power. The theme of this book is to recognize the mitochondrial role in fitness and to optimize mitochondrial function for performance and health.

Mitochondrial structure

Mitochondria are tiny. High-powered microscopy is needed to reveal their intricate structure. The small size and shape of mitochondria give a clue to their ancient bacterial origins. Like the cell, mitochondria have an outer membrane that regulates nutrient transport. There is also an inner membrane studded with the protein enzymes, which make ATP energy. Mitochondrial membranes restrain some of the harmful free radicals created during energy production, and mitochondrial antioxidant systems help neutralize free radicals [8]. The matrix is the jelly-like substance inside the mitochondrial inner membrane. The matrix contains the Krebs cycle enzymes that convert nutrients into the energized chemical intermediates that power the electron transport chain. Imbedded in the inner membrane, the electron transport chain pumps protons into the space between the inner and outer mitochondrial membranes. This acts as a tiny yet powerful proton battery that powers the enzyme that generates ATP (Figure 3.2). The matrix also contains mitochondrial DNA (mtDNA), which is more similar to bacterial DNA in structure and separate from the nuclear genome.

What is an enzyme?

An enzyme is protein (encoded by DNA) that performs a chemical reaction or function repeatedly. An enzyme is analogous to the auto factory worker who installs the hood. This worker is the best at what she does, performing this one function over and over again. Her work station has specialized tools, jigs, and robotic arms to facilitate the process. Similarly an enzyme performs the same singular function over and over again and has a specific three-dimensional structure to accomplish this task. When several enzymes are lined up as in the Krebs cycle and electron transport chain, a very efficient "assembly line" is created (Figure 3.3).

Mitochondrial DNA (mtDNA)

Although the maternal ovum and the paternal sperm both contain mitochondria, the relatively huge cytoplasm of the ovum contains many magnitudes more mitochondria than a sperm (about a 60,000:1 ratio in volume).[1] Mitochondria at the base of the sperm's

---

[1] A human ovum is 120–150 μm in diameter and a human sperm head is 5×3 μm in diameter. Volume = 4/3 πr^3.

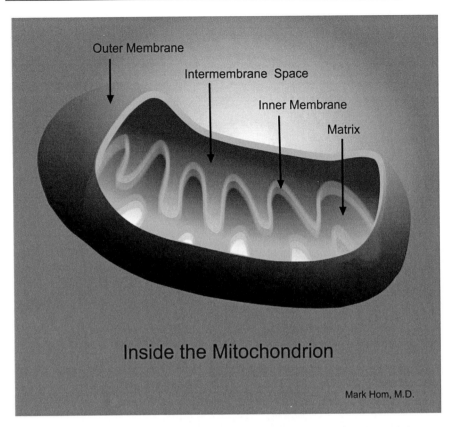

Figure 3.2 Dual membranes compartmentalize the mitochondrion. Fat burning and the Krebs cycle reside in the matrix, the electron transport chain lines the inner membrane, and the intermembrane space is the proton battery that powers ATP energy generation. Image by Mark Hom.

tail powers its swimming motion, but the sperm's mitochondria are vastly outnumbered by the ovum's mitochondria. It is theorized that the few paternal mitochondria which do enter the ovum are destroyed. Either through this destruction or simple dilution, we owe much of our mitochondrial genetics to our mothers (Figure 3.4). Over the many eons, many mitochondrial genes transferred to the central nucleus (where all other genes are stored) and are inherited in a normal Mendelian fashion with both parents making a contribution. However, some mitochondrial genes are passed on only by the mother via the ovum cytoplasm [9]. Although many professional sports focus on male prowess and select male winners, female genetics are even more important in athletic gift. With today's highly advanced *in vitro* fertilization techniques, it is possible to create offspring from three parents, with the third parent contributing the mitochondria. This has the promise of stopping transmission of defective maternal mitochondrial genes in affected families [10].

Figure 3.3 Assembly line workers and enzymes perform specific tasks.
Photography: General Motors.

Figure 3.4 The human ovum is about 60,000 times the size of a human sperm in volume.
Image by Mark Hom.

In the Peloton, there are few sons who have equaled or bettered the athletic results of their fathers. Taylor Phinney (son of Davis Phinney) is one exception, but his success may be attributed to his world champion mother's mtDNA (Connie Carpenter-Phinney). The successful brothers Frank and Andy Schleck share the same maternal mtDNA.

Mitochondrial origins and the endosymbiotic theory

The characteristics of bacteria-like size and shape, separate DNA, and bacteria-like chemistry led to the theory that the mitochondria in our cells are the result of an ancient infection, when mitochondria were engulfed by cells [11]. This occurred very early in evolution, when life was at the single cell stage. This collaboration was one of the biggest steps in evolution, as it gave cells enough energy to truly thrive rather than merely exist. This symbiosis worked so successfully that mitochondria became permanent passengers within our cells. We are therefore not pure beings, but are instead cohabitated conglomerations. When proposed in the 1920s, the endosymbiotic theory was rejected as being too bizarre [12]. Modern biology has since provided ample evidence to support this theory. For example, ribosomes are proteins associated with DNA function. Mitochondrial ribosomes are the same size as bacterial ribosomes and smaller than nuclear ribosomes. Antibiotics that interfere with bacterial ribosomes don't affect nuclear ribosomes, but can have side effects due to impairment of mitochondrial ribosomes.

Mitochondrial eve and the family tree

Unlike nuclear DNA which recombines with each generation, mtDNA is passed on intact (barring spontaneous mutation). Since the mitochondrial ancestral tree is undiluted with each generation, analysis of mtDNA facilitates chronological plotting of evolution. Female mitochondrial genes can be traced back to map human prehistoric migration patterns across the continents, and mtDNA studies have proven the Asia to the Americas land bridge theory [13]. Traced further back, mtDNA leads us to the origins of the first female human, our mitochondrial "Eve," in Africa [14]. Traced back even further, due to the slow changes in mtDNA over the eons, we find our place in the biological family tree that connects us to all life on the planet (Figure 3.5). Not to be left out, men have Y-chromosomal "Adam" genes, passed on from father to son [15].

Mitochondrial advantage in elite athletes

Differences in mitochondrial genetics and mitochondrial physiology are a suspected reason why some athletes seem more gifted and excel more than others. Elite athletes have not only superior mitochondrial physiology, but also the training regimen and willpower to boost their mitochondria to the maximum. Mitochondrial superiority may not mean super enzymes, but rather a lack of enzymatic defects and thus no weak links in the enzymatic chain of metabolic reactions. Subtle differences may explain why some athletes are sprinters, hill climbers, long-distance specialists or all-around athletes. Mitochondria have a role in determining the limits of athletic performance.

Despite their minuscule size, mitochondria contain complex enzyme systems. The Krebs cycle [16] occurs inside our mitochondria and extracts as much energy as possible from sugar, fat, and protein. All of these enzymes must work in a symphony to perform optimally. It is easy to see how one defective enzyme could interfere with the whole process. The Krebs cycle is often memorized by advanced biology students and young doctors because it is central to understanding metabolism (Figure 3.6).

The fact that mitochondria are central in explaining human power and athletic feats is reflected in popular culture. Consider the Star Wars explanation of the *Force*.

Star Wars: symbionts and the *Force*

George Lucas definitely knows his cell biology. In this dialog from the movie "Star Wars Episode I: The Phantom Menace," the source of the *Force* is revealed [17].

Anakin: "Master, Sir... I heard Yoda talking about midi-chlorians. I've been wondering: What are midi-chlorians?"

Qui-Gon: "Midi-chlorians are a microscopic life form that resides within all living cells."

Anakin: "They live inside me?"

Qui-Gon: "Inside your cells, yes. And we are symbionts with them."

Anakin: "Symbionts?"

Qui-Gon: "Life forms living together for mutual advantage. Without midi-chlorians, life could not exist and we would have no knowledge of the *Force*. They continually speak to us, telling us the will of the *Force*. When you learn to quiet your mind, you'll hear them speaking to you."

George Lucas uses the term "midi-chlorian" which is a play on the words "mitochondrion" (the singular form of mitochondria) and "chloroplast." The chloroplasts in plants are another example of endosymbiosis in biology. In the movie, Qui-Gon samples Anakin's blood and finds that his midi-chlorian levels are higher than even Yoda's, meaning Anakin is destined for great things. A high midi-chlorian count was a mechanism for Lucas to explain why some characters in the Star Wars saga were more naturally endowed with the *Force*. Mitochondrial genetics are likely a major reason why Greg LeMond is such a gifted athlete. Being the symbiotic dynamos in our cells, mitochondria are truly our life force. As to whether mitochondria speak to top athletes, there is this quote from one of the toughest riders in the Peloton, Jens Voigt: "Shut up, Legs!"

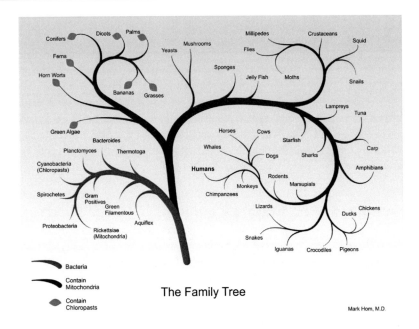

Figure 3.5 The gradual changes in mitochondrial DNA facilitate mapping of the phylogenetic family tree that connects all higher forms of life. Mitochondria and chloroplasts are endosymbionts that arose from ancient bacteria. Image by Mark Hom.

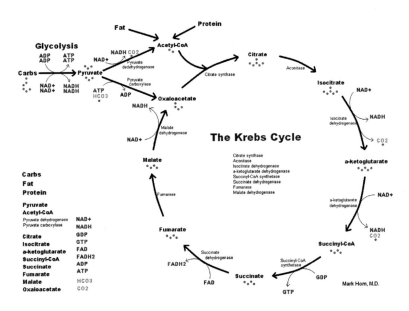

Figure 3.6 The Krebs cycle occurs inside our mitochondria. Image by Mark Hom.

Lucas got a few things wrong with the biology. The best way to sample mitochondria is with a muscle biopsy (as in Greg's diagnosis), not with a blood sample. Mature red bloods cells don't have mitochondria. Yoda dismisses pre-pubertal Anakin as being too old to start the Jedi training, when in fact significant mitochondrial boosting occurs in the teenage athlete during the hormone-induced growth spurt. In Lucas' world, the *Force* is inherited from the father. However, mitochondria are inherited maternally, not paternally. The famous and often misquoted line would have to be revised as, "Luke, I am your *Mother*," which would have a completely different ring to it! We'll give Lucas some literary license on that one. The original line from the movie is, "No, I am your Father," (Star Wars Episode V: The Empire Strikes Back) [18].

ATP synthase – the athlete's engine

The final step in the production of ATP involves the most intricate enzyme known in biology. Driven by the proton battery inside each mitochondrion, ATP synthase churns out ATP using a high-revving nanomotor mechanism [19]. ATP synthase is a complex of 31 proteins, with several parts analogous to the machinery in a high-performance automotive engine.

Powered by high-energy intermediates generated by the Krebs cycle, another enzyme system, the electron transport chain, pumps protons across the inner mitochondrial membrane into the intermembrane space, creating a proton gradient. The electron transport chain is driven by electron movement between enzymes. The very last electron accepter is oxygen (O_2), making water (H_2O). This tiny yet powerful proton battery drives the ATP synthase motor. The protons can only cross the mitochondrial inner membrane by entering a channel in membrane-bound ATP synthase. As protons enter this channel, they drive the base unit of ATP synthase (making it turn). This spinning can occur at very high speed (6000 rpm), as quickly as a high-performance automotive engine [20]. This conversion of fuel into electrical energy into proton battery power into mechanical motion is the way ATP is generated. The spinning base is rigidly connected to a central stalk, which is asymmetrically lobed like the camshaft of an automotive engine. Pierced by this stalk is a knob, shaped like a six-segmented orange, capable of holding three ADP (or ATP) molecules. Each 360° complete turn of the base and stalk generates three molecules of ATP. Spinning at high speed, each ATP synthase enzyme can produce tens of thousands of molecules of ATP per minute.

So if we were to use a microscope powerful enough to see inside living mitochondria, we would observe thousands of internal spinning nanomotors generating many billions of molecules of ATP. This spinning must occur at all times to sustain life and can be revved up to very high rpm when needed for intense exercise. The implications for athletes are many. Having strong, healthy, and non-leaky membranes is very important in maintaining the proton gradient that drives energy production. All of these enzymes are made of proteins requiring essential amino acids in the diet and healthy synthetic cell functioning. Genetic differences and genetic defects can explain athletic gift and inherited mitochondrial disorders. The number of mitochondria

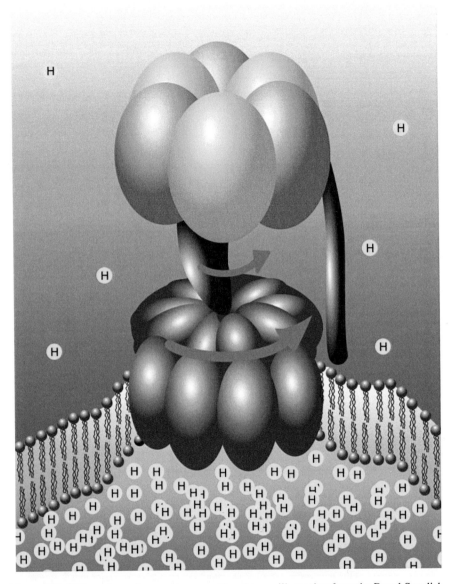

Figure 3.7 ATP synthase churns out energy. Based on an illustration from the Royal Swedish Academy of Sciences (1997). Press Release: The 1997 Nobel Prize in Chemistry. Image by Mark Hom.

can be increased with exercise, giving the well-trained athlete more ATP-generating nanomotors, and perhaps elite athletes have higher revving nanomotors than ordinary people. Variations of this enzyme give vitality to virtually every form of life on Earth. It is also the athlete's engine (Figure 3.7).

References

[1]   Raven PH, Johnson GB. Biology (updated version). 3rd ed Dubuque, Iowa: William C. Brown Publishers; 1995.
[2]   Grenfell JL, Rauer H, Selsis F, et al. Co-evolution of atmospheres, life, and climate. Astrobiology 2010;10(1):77–88.
[3]   Miller SL, Urey HC. Organic compound synthesis on the primitive Earth. Science 1959;130(3370):245–51.
[4]   Canfield DE. Oxygen: A Four Billion Year History. Princeton: Princeton University Press; 2014.
[5]   Yasuhito S, Katsuhiko S, Ryoko S, et al. Osmium evidence for synchronicity between a rise in atmospheric oxygen and Palaeoproterozoic deglaciation. Nature communications 2011;502:1–6.
[6]   Matsumi Y, Kawasaki M. Photolysis of atmospheric ozone in the ultraviolet region. Chemical Reviews 2003;103(12):4767–82.
[7]   Desqueyroux H, Pujet JC, Prosper M, et al. Effects of air pollution on adults with chronic obstructive pulmonary disease. Archives of Environmental Health 2002;57:554–6.
[8]   Cadenas E, Davies KJ. Mitochondrial free radical generation, oxidative stress, and aging. Free Radical Biology and Medicine 2000;29(3–4):222–30.
[9]   Sutovsky P, Moreno RD, Ramalho-Santos J, et al. Development: ubiquitin tag for sperm mitochondria. Nature 1999;402:371–2.
[10] Amato P, Tachibana M, Sparman M, Mitalipov S. Three-parent *in vitro* fertilization: gene replacement for the prevention of inherited mitochondrial diseases. Fertility and Sterility 2014;101(1):31–5.
[11] Thomas L. The Lives of a Cell: Notes of a Biology Watcher. New York: Bantam Books; 1974.
[12] Wallin IE. The mitochondria problem. The American Naturalist 1923;57(650):255–61.
[13] Fagundes NJR, Kanitz R, Eckert R, et al. Mitochondrial population genomics supports a single pre-clovis origin with a coastal route for the peopling of the Americas. American Journal of Human Genetics 2008;82(3):583–92.
[14] Cann RL, Stoneking M, Wilson AC. Mitochondrial DNA and human evolution. Nature 1987;325(6099):31–6.
[15] Mendez F, Krahn T, Schrack B, et al. An African American paternal lineage adds an extremely ancient root to the human Y chromosome phylogenetic tree. American Journal of Human Genetics 2013;92(3):454.
[16] Krebs HA, Weitzman PDJ. Krebs' citric acid cycle: half a century and still turning. London: Biochemical Society; 1987.
[17] Lucas G. Star Wars Episode I: The Phantom Menace. 20th Century Fox; 1999.
[18] Lucas G. Star Wars Episode V: The Empire Strikes Back. 20th Century Fox; 1980.
[19] Boyer PD. The ATP synthase – a splendid molecular machine. Annual Review in Biochemistry 1997;66:717–49.
[20] Stewart AG, Sobti M, Harvey RP, Stock D. Rotary ATPases: models, machine elements and technical specifications. Bioarchitecture 2013;3(1):2–12.

Feeding Your Cells

4

Nutrition and hydration

The previous chapters examined fitness and health at the organ system level, followed by delving deeper at the cellular and subcellular levels. That is because fitness is a whole body process and fitness is built one cell at a time. Now we will look even deeper, at the chemical level. Sports nutrition is the science of the macronutrient and micronutrient chemicals that rebuild our cells and provide our cells with energy. Macronutrients are the main components of the foods we eat: carbohydrates, protein, fat, hydration, and salts. These macronutrients and whole foods are the focus of this chapter. Micronutrients are the small yet essential vitamins and minerals in our food, which will be covered in Chapter 5: Mitochondrial Supplements.

People often liken food to the fuel in an automobile. Food is taken in and burned, with the residual passing out the tail pipe, so to speak. In a way that is true; both food and gasoline have caloric value, meaning they are chemical energy that can be turned into motion. However, the difference is that food becomes part of our human bodies. The caloric food we eat can be stored in the substance of our organs as glycogen and body fat. More importantly, food becomes incorporated into the individual cells in our bodies. Some cells in our body are constantly replacing themselves such as our skin cells, the lining of our gastrointestinal tract, and our blood cells; meaning that we are constantly building new cells with the building blocks that come from our food. Other cells are capable of growing, such as our muscles in response to resistance training. Other cells may not multiply or grow, but still need to remodel and repair themselves.

We *are* what we eat. Good or bad, whatever we choose to eat and drink becomes part of our bodies and cells. People often have a nihilistic approach to health, thinking it is genetic, socioeconomic, or because of other factors out of their control. But there are two ways we can intervene through choice: how much we exercise and what we include (or exclude) in our diet. Although it is easy to fall for trendy fad diets that focus on a single food group or single food item, what your body really needs is a varied and balanced diet. If you only concentrate on certain food groups at the exclusion of others, you may become deficient in important nutrients. Humans are omnivores, meaning that we have the digestive system and enzymes to utilize many types of food that no doubt served us well when we were foraging and adapting to various conditions during evolution. Because of this, we can survive on a wide range of different diets, from a strict vegetarian diet to a high animal protein diet; however, for optimal health, a balanced diet is best (Figure 4.1).

The Science of Fitness: Power, Performance, and Endurance. http://dx.doi.org/10.1016/B978-0-12-801023-5.00004-1

Figure 4.1 Fruits and vegetables supply nutrients, vitamins, carbohydrates, minerals, antioxidants, and fiber to the athlete's diet.
Image by Mark Hom.

Carbohydrates are the largest component of a normal diet and consist of sugars, starches, flour, grains, bread, potatoes, vegetables, and fruit. Carbohydrates are digested and then enter our blood stream as sugar to be burned as fuel or to be stored as glycogen. Not long ago, the main tenet of sports nutrition was carbohydrate loading: large amounts of carbohydrates were consumed in anticipation of athletic events. Although carbohydrates are a major source of our energy, sports nutrition is more than just eating a big bowl of pasta. We need other things in our diet to perform well: protein, good fats, hydration, salts, vitamins, and minerals. Carbohydrates are not only stored as glycogen (an energy reserve for intense exercise) but when eaten in excess, carbohydrates are also stored as body fat (for endurance energy). During our evolution, food was often scarce, so our bodies became very adept at storing fat during those rare times when ample food was available. The problem with current society is that food is now *always* available. There are good carbohydrates and not so good carbohydrates. The best carbohydrates come from whole foods because they contain vitamins, fiber, and protein. Refined and purified carbohydrates such as table sugar, high fructose corn syrup, white flour, and white rice are high in caloric value, but are "empty calories" that lack other nutrients. Ironically, there is a strong association between poverty and obesity [1]. Highly processed foods that are high in calories but low in nutrient value can be produced and marketed inexpensively, whereas fresh whole foods can be expensive to distribute, store, and purchase.

Although gluten-free foods are a recent trend, unless you have a documented gluten intolerance or sensitivity, a gluten-free diet is not generally recommended. Gluten is in so many products: nearly all breads, baked goods, condiments such as soy sauce, and as a thickening agent in many soups, that adhering to a gluten-free diet can be difficult. Gluten-free substitutes such as rice flour may have a higher glycemic index than whole wheat, and to make up for less flavor, often more calories in the form of refined sugars and fat are added.

Dietary protein is digested into amino acids, which are the building blocks of the synthesized protein in our bodies including: muscles, tendons, ligaments, structural connective tissue, and the matrix of our bones. There are 12 nonessential amino acids that we can produce internally. There are nine essential amino acids that we cannot produce internally and therefore must be included in our diet.

Essential Amino Acids (Must be included in the diet)	Non-Essential Amino Acids (Made by the body)
Histidine	Alanine
Isoleucine	Arginine
Leucine	Aspartic acid
Lysine	Cysteine
Methionine	Glutamic acid
Phenylalanine	Glutamine
Threonine	Glycine
Tryptophan	Proline
Valine	Serine
	Tyrosine
	Asparagine
	Selenocysteine

Fish, red meat, dairy product, eggs, soy, beans, and nuts are all sources of essential and nonessential amino acids. For those on strict vegetarian diets, consultation with a dietician is recommend to make sure that the correct amounts and combinations of nonanimal protein are eaten to account for all of the amino acids your cells need to make protein. A current fad is that athletes need lots of protein with every meal because protein builds muscle. However, there is only so much muscle building that takes place at any given time, and excess protein can contribute to body fat or must be eliminated by the kidneys. A high-protein diet may accelerate the decline in renal function in people with mild renal insufficiency [2]. A diet high in meat can also include steroids, growth hormones, and antibiotics that create more yield in the livestock industry [3]. It can also result in too many calories, too much saturated fat, too much salt, and too much iron for some people. Too much intake of large, long-lived fish can increase exposure to mercury and farm raised fish can have higher levels of toxins [4]. Non-meat sources of protein such as soy protein power, whey, nuts, eggs, yogurt, and beans are good alternative sources of protein. Although protein powders are a convenient way to supplement protein, more is not better and moderation is recommended.

Dietary fat is often blamed for obesity and poor health. The modern view of dietary fat has changed. Although it is true that fat has high caloric density, it is not as easily absorbed as other food types and dietary fat is not the only food group that contributes to body fat. Excess carbohydrates and excess protein can also be stored as body fat, so it is often the nonfat ingredients in the diet that contribute to obesity. Fats are a major aspect of our energy storage metabolism, being the main source of endurance energy and for survival in times of starvation (while helping to preserve muscle). Fat molecules are an essential part of our cell membranes and many vital hormones. The most important modern concept is that all dietary fats are not the same; there are bad fats and good fats. The worst types of fats are trans fatty acids (trans fat for short) because they are chemically altered and unnatural. Our bodies take in and incorporate trans fats but they disrupt normal cell function and metabolism. Trans fats change the fluidity of cell membranes, raise bad cholesterol, lower good cholesterol, accelerate arteriosclerosis, and are associated with increased cancer risk [5]. Ways to avoid trans fats are described later in this chapter and in Chapter 8: When Things Go Wrong (Mitochondrial Toxins section). Saturated fat from animal sources, dairy products, and tropical oils was for many decades blamed for obesity and heart disease causing a shift toward low-fat diets and more carbohydrate intake. However, obesity during this time skyrocketed and heart disease did not go away. Saturated fats may not be as bad as once thought, since they are naturally occurring and our bodies know how to deal with them. Still, many heart organizations regard saturated fat as a risk factor for developing cardiovascular disease and recommend limits and moderation in saturated fats consumption [6]. Good fats include Omega-3 fats (found in oily fish) and mono-unsaturated fat (in olive oil and avocados), which are beneficial to cell function, can prevent arteriosclerosis, reduce inflammation, improve insulin response at the cell membrane level, and improve longevity.

Fighting trans fat at the institutional level

At the hospital where I work, I often skip the meat entrée and order four side vegetables for lunch. One day I noticed that the vegetables had a more buttery flavor than usual. I asked to speak with the cafeteria manager who said that they ran out of the usual oil for steaming the vegetables and substituted their deep frying oil. When I asked if they were using a healthy frying oil he said "Yes!" and went to retrieve the bottle. The nutritional facts label revealed that the main component was partially hydrogenated vegetable oil. Although "partially," "hydrogenated," and "vegetable oil" sound innocuous, this is the code phrase for trans fat. I found an article that said that serving trans fat at a hospital cafeteria was the equivalent of having ashtrays in the operating room (the opposite of healthy) [7]. I did not expect a sudden change but the very next morning the cafeteria manager emailed me and said that their cooking and frying was now completely trans fat free. Converting was as simple as checking a different box on their inventory supply list. With that simple change, thousands of patients, employees, and visitors now have healthier cafeteria food.

Mark Hom

Hydration and salt balance is a key part of physical performance during strenuous exercise. Our muscles and organs need to be perfused with an adequate blood volume and individual cells need hydration to function at peak efficiency. Strenuous aerobic activity generates heat, which must be dissipated with sweat to stay within the narrow range of optimal temperature. Sweating causes loss of water and salt that must be replaced. The reasons why sports drinks improve hydration will be described later in this chapter.

Healthy and unhealthy diets around the world

There are several regional tradition diets that are known to be healthy. The native Greenland Eskimo diet consists largely of salmon, seal meat, and seal blubber (seals eat salmon) and the Eskimo incidence of heart disease is very low. This was the major clue that omega-3 fish oils were beneficial to health [8]. When these Eskimos eat a typical western diet, they have the same high incidence of heart disease as the modern world, so they are not genetically predisposed to lower heart disease risk.

Okinawans have some of the highest longevity statistics in the world. The traditional Okinawan diet is high in vegetables, soy, and legumes, and low in red meat. In most of Japan, the main staple food is rice, but in Okinawa it is sweet potato.

The French have a high fat content diet but have lower heart disease risk and lower obesity rates compared with the American diet. The French eat more good fats and less bad fats, have very low trans fat intake, eat more fresh food and less processed food, consume smaller portions, and engage in more baseline physical activity (more walking and less driving) [9]. The also drink more red wine; however, the benefits of the antioxidant resveratrol are unlikely to be the major cause of the "French Paradox."

The typical American diet is one of the least healthy with high rates of cardiovascular disease and obesity. The stereotypical American diet is high in refined and processed foods, high in calories, low in nutritional value, high in added sugar and salt, low in fiber, high in fried foods, high in trans fat, geared for prepackaged convenience rather than freshness, and imbalanced toward meat instead of fruits and vegetables.

Recently, Mexico surpassed the US for having the highest obesity rate (32.8 percent) in the world [10]. The traditional Mexicans diet has shifted away from whole grains and vegetables to more refined grains and meat. Mexican consumption habits have shifted toward a more American cultural model. The physical activity of farming has been greatly replaced with mechanization. And high levels of poverty have resulted in a diet that is rich in calories and poor in nutritional value.

The Mediterranean diet

Large multi-national studies have compared the typical western/American diet to the Mediterranean diet, which is proving to be one of the healthiest diets in the world [11,12]. The study confirmed suspicions that the typical American diet was very high risk for cardiovascular disease and that the Mediterranean diet was protective against heart disease. In fact, the results showed such a dramatic difference in cardiac events and overall mortality that one such study was terminated early. It was deemed unethical to keep half of the study population on the dangerous American diet [13].

Health benefits of the Mediterranean diet

- People living in the areas around the Mediterranean Sea have low risks of developing heart disease, cancer, and other diet-related illnesses.
- These people also had one the highest adult life expectancies in the world.
- Following a traditional Mediterranean diet, engaging in regular exercise, and not smoking can prevent more than 80 percent of new cases of heart disease, 70 percent of stroke, and 90 percent of type 2 diabetes.
- May be associated with a slower onset of cognitive decline.
- Reduces the activity of genes involved in atherosclerosis.
- Limits production of inflammatory chemicals.
- Reduces formation of abnormal blood clots.
- 10–20 percent less likely to die from heart disease, cancer, or any other cause over the course of the study.
- Those having a previous heart attack were significantly less likely to have a second heart attack.
- Lower risk of developing type 2 diabetes.
- Lost more weight and showed greater improvement in risk factors for heart disease.
- The protective effects were so striking (a 70 percent reduction in all causes of death) that one study was terminated after 27 months (rather than the planned 5 years).

Ingredients of the Mediterranean diet include
- Low red meat
- Non-red meat sources of protein (nuts, beans, fish)
- Fish oil (omega-3 fat)
- Olive oil (monounsaturated fat)
- Whole grains and fiber (low refined sugars)
- Red wine (resveratrol)
- Fresh fruits and vegetables (antioxidants and fiber)
- Tomato sauce (concentrated lycopene)
- Variety and freshness of food

Superfoods
Although we recommend a varied and balanced diet, there are some foods that are especially good for you and should be included or increased in your diet.

Sardines are an example of a small, wild, oily fish that is rich in healthy omega-3 oil. Small short lived fish are lower on the food chain and accumulate fewer toxins. Farm-raised salmon raised on fish meal tend to be fatter with more beneficial omega-3 content than wild salmon, but unfortunately farm-raised fish can also have more toxins such a polychlorinated biphenyls and dioxins [14]. Large fish such as tilefish from the Gulf of Mexico, shark, swordfish, and king mackerel can accumulate high levels of mercury, so they should be avoided by women of childbearing age to reduce fetal mercury exposure, and white (albacore) tuna intake should be limited by these women [15].

Whole grains contain the bran (the outer skin containing B vitamins and fiber) and the germ (the sprout embryo containing B vitamins, protein, minerals, and healthy fats). Refined white flour only has the endosperm (the starchy majority of the grain). Be aware that some manufacturers add brown food coloring (caramel) to white flour products to make them appear to be whole grain. The best way to recognize real whole grain bread is by its coarser texture (due to the fiber) [16]. This is another example where you must read the food label carefully.

Eggs contain all of the nutrients and protein a chick needs to grow and survive for 21 days before it hatches, so eggs are very nutritionally dense. They contain all of the essential amino acids and are especially rich in the muscle-building essential amino acid, leucine [17].

Steak (especially grass-fed beef) is nutritionally dense with protein, iron (in heme form that is more readily absorbed), B vitamins (including high amounts of B_{12}), and minerals. A steak can be chosen for leanness and quality. Eating red meat every day is not necessary or recommended. One way to include red meat is to eat it once a week. You get the nutritional benefit of the protein, vitamins, and iron but you will not accumulate as much of the saturated fat, cholesterol, and livestock enhancers (antibiotics and hormones). You can eat like a vegetarian the rest of the week, but not deny yourself on the one day of the week when you eat red meat. A high red meat diet increases the risk of colon cancer [18]. A low-red-meat diet can be perfectly healthy. Grass-fed beef is more nutritious than grain-fed beef because there are higher levels of good omega-3 fat and good conjugated linoleic acid (CLA) in grass-fed cattle [19].

Yogurt (with active cultures) is rich in calcium, protein, vitamin B_2, vitamin B_{12}, vitamin D (when fortified), potassium, and magnesium. Active culture yogurt contains probiotic "good" bacteria that help digestion, immunity, and reduce obesity [20]. The healthiest yogurt flavor to buy is plain with no added sugar.

Nuts are an excellent source of protein, omega-3 fat, vitamin B_2 (riboflavin), vitamin E, iron, zinc, and magnesium. They are low in cholesterol and sodium (if unsalted), and have a very low glycemic index. Mixed with dried fruit or raisins, they are a healthy and satisfying snack.

Soybeans are high in protein, omega-3 fat, calcium, vitamin B_9 (folate), and vitamin B_1 (thiamine). Tofu is used as an effective meat substitute in many vegetarian eastern cultures. The reason why this legume is high in protein is because symbiotic bacteria living in the root nodules of the soybean plant can capture nitrogen (N_2) from the air for amino acid synthesis. This also enriches the soil with nitrogen and is the reason why soybeans are used in crop rotation because they naturally fertilize the soil and improve our food supply. It is questionable whether soy can lessen menopausal symptoms in women [21]; it does not have a feminizing effect on men [22].

Oatmeal is high in fiber, has more protein than most grains, moderates carbohydrate absorption, lowers bad low-density lipoprotein cholesterol, and can decrease cardiac disease risk. A heart healthy diet including oatmeal, psyllium fiber, soy protein, and low animal fat can be just as effective in lowering cholesterol as first generation statin medications [23].

Cruciferous vegetables such as broccoli, Brussel sprouts, cabbage, kale, and bok choy are high in fiber, vitamin A, B vitamins, vitamin C, vitamin K, magnesium, antioxidants, and sulfur compounds that can help your liver detoxify your system [24].

Dark berries are high in phytoflavinoids (powerful antioxidants), vitamin C, potassium, and polyphenols that may have benefits in anti-aging and brain function [25].

Beets are high in fiber, antioxidants, vitamin B9 (folate), manganese, magnesium, potassium, vitamin C, and vitamin B_6 (pyridoxine). Drinking large amounts of beet juice can increase endurance, possibly due to a nitrate effect of increasing blood flow to muscles [26].

Avocados have a creamy texture and flavor due to their high fat content. This is an example of a high fat food that is good for you because it is a good fat (monounsaturated). They are also high in protein, fiber, potassium, vitamin K, vitamin B_9, vitamin B_6, vitamin B_5, vitamin C, and vitamin E (Figure 4.2).

Foods to avoid
Perhaps the easiest way to improve your diet is to exclude unhealthy food items. Although some of them are created to pacify craving, you soon forget about them once they are replaced with healthy and good tasting alternatives.

Figure 4.2 Eating healthy does not mean having to sacrifice flavor. Strawberry and avocado salad drawn from a photograph by Kevin Lynch at www.closetcooking.com. Image by Mark Hom.

Deep fried foods are tempting due to the aroma, crispy outer texture, moist inner texture, and added salt. But if you compare a batter-dipped deep-fried chicken breast with its skin to a grilled skinless chicken breast, you are getting twice the calories for the same amount of protein. Is the crispy batter (which soaks up the deep frying oil) and the greasy skin (saturated fat) really worth the few minutes of gratification? You could be carrying molecules of that fried chicken meal in your cells for weeks. If you eat at a restaurant, you will not be able to know with certainty that they do not use unhealthy trans fat frying oil. When breading (deep fried food) and potatoes (French fries and potato chips) are fried at high temperature, an amino acid (asparagine) combines with sugars to form the chemical acrylamide. Acrylamide has been called a "probable human carcinogen' by the National Toxicology Program [27] and poses a "major concern" according to the World Health Organization [28]. If you stop eating deep fried food, you can dramatically lower caloric intake, lower trans fat intake, and decrease exposure to a probable carcinogen.

Prepackaged snacks are moist, sweet, salty, and convenient. But they may have trans fat, excess sugar, too many calories, and too much salt. Most nutritionists agree that there is no safe level of added dietary trans fat and it should be removed from the diet [29]. Due to a food packaging loop hole the nutrition facts label may hide the fact that the item has trans fat. If the serving size (often unrealistically small) contains less than 0.5 g of trans fat, the manufacturer is allowed to claim 0 g trans fat. To make sure, you must look at the ingredient list, which (by the way) will not say trans fat (another deception). It will say partially hydrogenated vegetable oil, the code phrase for trans fat.

Cake can be loaded with trans fat in the icing and in the moist dough, and cake is often presented for consumption without a nutrition facts label. The calorie count can be very high and it is often eaten after a large meal when hunger is already satisfied.

Soft drinks have been linked to obesity, type 2 diabetes, kidney damage, and pancreatic cancer [30,31]. Regular soft drinks can contain huge amounts of empty calories with either pure sugar or high fructose corn syrup. Diet sodas may be even worse, triggering hunger and resulting in even more weight gain than regular sodas. The phosphoric acid in colas can be harmful in people with impaired renal function and can weaken bones by disrupting the calcium phosphorus balance [32].

Ground beef can be low quality, can contain additives, and is a commonly recalled food item due to bacterial contamination. Although red meat does have some good nutrients, this is another example of processed food being less safe and less healthy. Whenever there is a recall of tons of beef for bacteria, it is always ground beef and never filet mignon. When photographs of extruded lean finely textured beef (LFTB) nicknamed "pink slime" were leaked to the media, there was public outrage [33]. LFTB is made of low-quality beef trimmings and fat (which would otherwise be discarded) that is treated with ammonium hydroxide and added to ground beef. At one time LFTB "pink slime" was in 70 percent of US ground beef. Now that the uproar has abated, there are likely plans to bring it back because it is still Food and Drug Administration approved and does not even require food labeling.

Hydration

The human body is composed of 60–70 percent water. Our cells are tiny droplets of water contained within thin membranes. The organelles (such as our mitochondria) and the vital enzyme systems inside our cells all work in an aqueous environment. Strenuous exercise and various other conditions can lead to dehydration. Therefore, it is important for athletes to understand hydration and to be well hydrated for optimal physical performance.

One way to consider hydration is to view the human body as a lake or reservoir. For a recreational lake to remain full and host swimmers, boaters, and wildlife, there has to be a balance between water inflow and outflow. If water is lost through outflow from the dam or evaporation on a hot day, the reservoir's water level can remain at normal levels as long as there is adequate inflow. Activity on the lake can remain at high levels. However, the reservoir level may fall if there is either too much outflow or inadequate inflow. If the water level drops too low, then activity on the lake can slow, or in a drought stop completely. By understanding a few principles in this section, you can remain well hydrated and perform optimally in all conditions.

In our analogy, the lake is not fresh water, it is salt water. Our blood, tissues, and cells are saline (salts dissolved in water). In fact, the salinity of our bodies is very close to the salinity of sea water, where our ancient ancestors originated. Outside the cells, the main salt is sodium (Na^+) and inside the cells, the main salt is potassium (K^+). As noted in Chapter 2, this difference in Na^+ and K^+ concentrations is maintained by tiny ATP-powered pumps in our cell membranes, which in total use the most amount of ATP energy in the body [34]. Therefore, hydration is more than just water; it is also about salt and energy.

Technically, the term "dehydration" means loss of water and does not say much about the salt losses. Medically, a more accurate term is volume depletion (loss of fluid from the circulatory system). If you are ill with volume depletion your vital signs, physical exam, history, and blood tests can reveal if you are low on water, salt, or both. The doctor can then decide how to best replace your fluids and whether the fluids can be replaced with drinking or if you might need an intravenous line (IV) for more rapid fluid replacement. But the term athletes are most familiar with is dehydration, which is the term we will use. Because our hydration status is so important, is it closely regulated by your autonomic nervous system, pituitary gland, brain (sense of thirst), kidneys, and hormone control (antidiuretic hormone and aldosterone). This complex system is very good at conserving water and releasing excess water when required. However, there are conditions and situations where this regulation cannot keep up and we can become dehydrated. Large amounts of water and salt can be lost with vomiting and diarrhea, which is a major cause of childhood death in developing countries [35]. Water and sugar can be lost through the urine in diabetes which is suspected when patients have a history of extreme thirst and frequent urination.

However, for the athlete the main concern is to avoid dehydration when exercising. We lose water from perspiration, urination, defecation, and respiration (water vapor in the breath). We gain water from eating, drinking, and a small amount produced from our mitochondria during metabolism. Just as with our lake analogy, we are always losing some water that needs to be replaced. We even lose some water when we sleep, from breathing and urination (and we obviously do not drink when we are asleep). When we lie down, the blood pressure going to our kidneys decreases and urine going to our bladder slows down but does not stop completely. It is a good idea to go to bed well-hydrated to account for these nighttime losses. If you have a big training day or race in the morning, you do not want to start the day with a low reservoir. Similarly you may want to drink a tall glass of water when you first wake up. Although we may need 2 L of water per day, much of that comes from food and does not have to come from a bottle or glass. Furthermore, the amount of water an individual needs depends on their body size, muscle mass, activity level, exposure to the elements, and climate conditions. A better gauge may be urination. You should drink enough so you need to urinate at least three to four times a day and the urine should be light yellow in color (concentrated dark urine can indicate dehydration). You can also drink too much water. If people drink massive amounts of plain water and not enough salt, the sodium level in the blood stream can drop dangerously low (hyponatremia) and in severe cases may lead to brain swelling, seizures, and death [36]. So although water is a great way to keep hydrated and stay cool, it should be used sensibly and in moderation.

It is a good idea to drink some fluids just before a training session or event. The body starts losing water very quickly with strenuous exercise and pre-event hydration will top off the tank and will be one less bottle you need to carry. During an event it is important to keep drinking even before you become thirsty. Sweating from a hard workout, especially in hot weather will cause a steady loss of water and salt. Rapid breathing will also create water vapor loss.

Sports drinks are better at fluid replacement than plain water because they contain salt and sugar. The salt in sports drinks not only replaces the salt lost with sweat, it increases the rate of water absorption from the intestines, and helps retain the water by slowing urination of water. Although athletes may think the sugar in sports drinks is a flavoring or for fuel, it is really to help speed the absorption of water. When one glucose molecule is actively transported from the intestines to the blood stream, it brings with it six molecules of H_2O. When exercising all day it is important to stop sometimes and rehydrate; that way you can perform at your best. Sports drinks speed the rehydration process. Rehydration is especially important in hot weather to prevent heat exhaustion and heat stroke. Go back to Chapter 2 (Homeostasis section) to review the signs and symptoms of overheating so you can prevent this from happening.

On a bicycle, it is often a good idea to carry plain water. It is useful for splashing on your body to create instant evaporative cooling on a hot day. It is good for washing your face or removing dust from your eyes. And it is still a good way to rehydrate while being less sticky and messy on your equipment. For a short but intense ride, you may only need to carry plain water. In cool weather, you may not have as much thirst, but you are still losing water from sweating so do not neglect hydration when it is cooler.

After an intense exercise session and despite trying to maintain your hydration, you are probably underhydrated and possibly dehydrated, so it is important to drink plenty of fluids after exercise. This will help you recover and clear metabolites from your system, and also prepare you for the next day by keeping your reservoir full.

References

[1] Levine JA. Poverty and obesity in the U.S. Diabetes 2011;60(11):2667–8.
[2] Knight EL, Stampfer MJ, Hankinson SE, et al. The impact of protein intake on renal function decline in women with normal renal function or mild renal insufficiency. Annals of Internal Medicine 2003;138(6):460–7.
[3] Jeong SH, Kang D, Lim MW, et al. Risk assessment of growth hormones and antimicrobial residues in meat. Toxicology Research 2010;26(4):301–13.

[4] Scheer R, Moss D. Harvest of fears: farm-raised fish may not be free of mercury and other pollutants. Scientific American June 15, 2011; www.ScientificAmerican.com.

[5] Dhaka V, Gulia N, Ahlawat KS, et al. Trans fats—sources, health risks and alternative approach—A review. Journal of Food Science and Technology 2011;48(5):534–41.

[6] German JB, Dillard CJ. Saturated fats: what dietary intake? American Society for Clinical Nutrition. American Journal of Clinical Nutrition 2004;80(3):550–9.

[7] Jacobson MF. Top hospitals harming hearts by cooking with trans fat. *Center for Science in the Public Interest* February 6, 2006; www.cspinet.org.

[8] O'Keefe JH Jr, Harris WS. From Inuit to implementation: omega-3 fatty acids come of age. Mayo Clinic Proceedings 2000;75(6):607–14.

[9] Ferrières J. The French paradox: lessons for other countries. Heart 2004;90(1):107–11.

[10] Daniel J. How Mexico got so fat and is now more obese than America. Mail Online News 2013; www.dailymail.co.uk.

[11] Estruch E, Ros E, Salas-Salvadó J, et al. Primary prevention of cardiovascular disease with a Mediterranean diet. New England Journal of Medicine 2013;368:1279–90.

[12] de Lorgeril M, Salen P, Martin JL, et al. Mediterranean diet, traditional risk factors, and the rate of cardiovascular complications after myocardial infarction: final report of the Lyon Diet Heart Study. Circulation 1999;99:779–85.

[13] Leaf A. Editorials: dietary prevention of coronary heart disease, the Lyon Diet Heart Study. Circulation 1999;99:733–5.

[14] Hamilton MC, Hites RA, Schwager SJ, et al. Lipid composition and contaminants in farmed and wild salmon. Environmental Science & Technology 2005;39(22):8622–9.

[15] U.S. Food and Drug Administration. Fish: what pregnant women and parents should know. Draft Updated Advice by FDA and EPA, www.fda.gov; 2014.

[16] CBS News. White bread in wheat bread's clothing. February 8, 2008; www.CBSNews.com.

[17] Norton LE, Wilson GJ, Layman DK, et al. Leucine content of dietary proteins is a determinant of postprandial skeletal muscle protein synthesis in adult rats. Nutrition & Metabolism (London) 2012;9(1):67.

[18] Chan DS, Lau R, Aune D, et al. Red and processed meat and colorectal cancer incidence: meta-analysis of prospective studies. PLoS One (Public Library of Science) 2011;6(6):e20456.

[19] Ponnampalam EN, Mann NJ, Sinclair AJ. Effect of feeding systems on omega-3 fatty acids, conjugated linoleic acid and trans fatty acids in Australian beef cuts: potential impact on human health. Asia Pacific Journal of Clinical Nutrition 2006;15(1):21–9.

[20] Yadav H, Lee JH, Lloyd J, et al. Beneficial metabolic effects of a probiotic via butyrate induced GLP-1 secretion. Journal of Biological Chemistry 2013;288:25088–97.

[21] Levis S, Strickman-Stein N, Ganjei-Azar P, et al. Soy isoflavones in the prevention of menopausal bone loss and menopausal symptoms: a randomized, double-blind trial. Archives of Internal Medicine 2011;171(15):1363–9.

[22] Messina M. Soybean isoflavone exposure does not have feminizing effects on men: a critical examination of the clinical evidence. Fertility and Sterility 2010;93(7):2095–104.

[23] Jenkins DJ, Kendall CW, Marchie A, et al. Direct comparison of a dietary portfolio of cholesterol-lowering foods with a statin in hypercholesterolemic participants. American Journal of Clinical Nutrition 2005;81(2):380–7.

[24] Percival M. Phytonutrients & detoxification. Clinical Nutrition Insights, Foundation for the Advancement of Nutritional Education 1997;5(2).

[25] Poulose S et al. Eating berries may activate the brain's natural housekeeper for healthy aging. 240th National Meeting of the American Chemical Society, USDA; 2010.

[26] Lansley KE, Winyard PG, Bailey SJ, et al. Acute dietary nitrate supplementation improves cycling time trial performance. Medicine & Science in Sports & Exercise 2011;43(6):1125.

[27] Fuhr U, Boettcher M, Kinzig-Schippers M, et al. Toxicokinetics of acrylamide in humans after ingestion of a defined dose in a test meal to improve risk assessment for acrylamide carcinogenicity. Cancer Epidemiology Biomarkers and Prevention 2006;15(2):266–71.

[28] World Health Organization. Summary report of the sixty-fourth meeting of the Joint FAO/WHO Expert Committee on Food Additives (JECFA). Food and Agriculture Organization of the United Nations 2008.

[29] National Academy of Sciences. Dietary Reference Intakes for Energy, Carbohydrate, Fiber, Fat, Fatty Acids, Cholesterol, Protein, and Amino Acids (Macronutrients). Food and Nutrition Board, Institute of Medicine of the National Academies, *National Academies Press*, 424; 2005.

[30] Apovian CM. Sugar-sweetened soft drinks, obesity, and type 2 diabetes. Journal of the American Medical Association 2004;292(8):978–9.

[31] Mueller NT, Odegaard A, Anderson K, et al. Soft drink and juice consumption and risk of pancreatic cancer: the Singapore Chinese Health Study. Cancer Epidemiology, Biomarkers & Prevention 2010;19:44.

[32] Tucker KL, Morita K, Qiao N, et al. Colas, but not other carbonated beverages, are associated with low bone mineral density in older women: The Framingham Osteoporosis Study. American Journal of Clinical Nutrition 2006;84(4):936–42.

[33] CBS News Staff. "Pink slime" outcry causes Beef Products Inc. to close three plants. CBS News May 10, 2012; www.CBSNews.com.

[34] Skou JC, Esmann M. The Na, K-ATPase. Journal of Bioenergetics and Biomembranes 1992;24:249–61.

[35] Wardlaw T, Salama P, Brocklehurst C, et al. Diarrhoea: why children are still dying and what can be done. Lancet 2010;375(9718):870–2.

[36] Moritz ML, Ayus JC. The pathophysiology and treatment of hyponatraemic encephalopathy: An update. Nephrology Dialysis Transplantation 2003;18(12):2486–91.

Mitochondrial Supplements

Vitamins and nutritional supplements

In this section, we will review (with a skeptical eye) the scientific evidence regarding supplements that are claimed to support mitochondria and energy metabolism. Although the following supplements are natural and generally considered safe in recommended amounts, they should only be taken with the approval of your physician, especially if you are taking other medications or have allergies. Note that, the only pill Greg took when he raced was a multivitamin with iron to supplement a balanced diet. We are not suggesting that supplements are required to be physically fit or to win races. These supplements are not intended to replace exercise. Nor should they be taken primarily for weight loss. There is no substitute for physical exercise and there is no such thing as exercise in a bottle. The supplements we will review include vitamins, coenzyme Q10, acetyl-L-carnitine, R-lipoic acid, magnesium (Mg), resveratrol, fish oil, olive oil, multivitamins with iron, and dietary fiber.

What is a vitamin?

There are certain organic compounds that we require to thrive. These compounds may be made in the body but are produced in too small of a quantity and therefore must be included in our diet. Vitamin requirements depend on the species. For example, humans require vitamin C because we do not make enough, whereas dogs do not need vitamin C because they make enough for their needs. When people think about vitamins they probably think about pills in a bottle, but our food contains ample amounts of vitamins. Vitamin deficiencies occur when people have poor access to fresh food or extremely limited variety in their diet. For example, beriberi is the disease that occurs when there is not enough vitamin B_1 (thiamine) in the diet. Beriberi can be seen in cultures where the main food source is dehusked white rice (the husk contains thiamine). Pellagra can be seen in extreme poverty, alcoholism, and a mainly corn-based diet due to vitamin B_3 (niacin) deficiency.

The most well-known example of vitamin deficiency is scurvy. During the age of exploration, it was noted that seafarers (who had limited access to fresh foods) developed a constellation of symptoms including lethargy, shortness of breath, muscle and bone aches, rough skin, loose teeth, and poor wound healing (quite the opposite of a healthy and vibrant sailor). One clue was that scurvy affected the crew (who ate basic rations) more than the officers who enjoyed a more varied diet. James Lind, a ship's surgeon with the British Royal Navy, found that scurvy could be prevented and reversed with consumption of citrus fruit (vitamin C) and is why British sailors were nicknamed "Limeys."

In modern times, we are highly unlikely to have a severely restricted diet. We no longer subsist on what is grown on our small piece of property or what can be kept in

The Science of Fitness: Power, Performance, and Endurance. http://dx.doi.org/10.1016/B978-0-12-801023-5.00005-3

dry storage. Now our food is transported from all over the world, traveling hundreds if not thousands of miles to reach our plate. Therefore, we are highly unlikely to suffer the vitamin deficiency diseases of yore. For healthy people who eat a balanced diet, there is little or no evidence that vitamin supplements make us any healthier. Ironically, the people who choose to take a daily multivitamin generally are concerned enough about their health that they already eat a balanced diet and therefore do not need vitamin supplementation [1]. Just because correcting vitamin deficiency is good for you, taking more (mega doses of vitamins) is not necessarily better. When certain vitamins are taken in excess they can be toxic or have negative health effects. Vitamins A, D, E, and K are fat-soluble vitamins that can build up in the body. Excessive amounts of some vitamins such as A, E, and K can have negative effects on health.

On the other hand, there are reasons why a daily multivitamin should be considered. Our modern diet often contains overly refined foods that have had vitamins removed during processing. Overly cooked cafeteria-style vegetables can have reduced vitamin content. There are some people who need extra vitamins such as: pregnant or breast-feeding women, chronic alcoholics, extreme dieters (ill-advised), people on restricted diets due to food allergies, the elderly, some vegetarians, and those with malabsorption disorders. There are some processes in the body that require multiple complex steps, one example being the production of red blood cells (RBCs). Hemoglobin synthesis and RBC production is a complicated multistep process that requires more than just iron; it also requires vitamin B_5, vitamin B_6, vitamin B_9, vitamin B_{12}, vitamin C, vitamin E, Mg, and zinc. Athletes should think about getting their vitamins from a well-balanced and varied diet and avoid mega vitamin doses. A daily multivitamin pill contains safe levels of vitamins and is formulated to cover all of your bases, preventing potential gaps in nutrition.

B complex vitamins are often called the energy vitamins because they are important for metabolism and help your mitochondria convert food into energy. These complex mitochondrial functions include: the Krebs cycle (converting food into energy), the electron transport chain (generation of ATP energy), and beta oxidation (fat burning). Several of the B vitamins also help with hemoglobin and RBC production, which helps keep our organs and mitochondria well oxygenated.

The B vitamins include:

- Vitamin B_1 (thiamin) helps produce acetyl coenzyme A (energy source of the Krebs cycle).
- Vitamin B_2 (riboflavin) helps with the electron transport chain, Krebs cycle, and fat burning.
- Vitamin B_3 (niacin) conveys energy from the Krebs cycle to the electron transport chain.
- Vitamin B_5 (pantothenic acid) helps burn fat and carbohydrates.
- Vitamin B_6 (pyridoxine) helps with hemoglobin synthesis and gluconeogenesis.
- Vitamin B_7 (biotin) helps with multiple metabolic processes.
- Vitamin B_9 (folate) helps RBC production and prevents neural tube birth defects.
- Vitamin B_{12} (cobalamins) helps RBC production and nerve sheaths.

Vitamin C has antioxidant properties, helps build collagen (connective tissue and wound healing), helps immunity, helps improve cholesterol levels, and may help slow atherosclerosis. In the old days, sailors with scurvy were the exact opposite of healthy: weak, tired, achy, and with mouth and skin sores. Vitamin C is water soluble and is not retained for long in the body, so we need to replenish vitamin C. Fortunately, vitamin C is abundant in a healthy diet that is rich in fresh fruits and vegetables. Vitamin C helps us absorb iron. Heme iron (present in red meat and blood) is readily absorbed, but non-heme iron (that might be taken in supplementary pill form) is harder to absorb. Vitamin C helps people who need to take iron supplements by assisting the absorption of non-heme iron [2].

Vitamin D is an example of a vitamin that we produce in the body, but under certain conditions the amount is not enough. Vitamin D synthesis is rather complicated, taking place in multiple organs including the skin, liver, and kidneys, with skin sunlight exposure a key step. One major function of vitamin D is in bone health. Our bones are not solid unchanging blocks of calcium, but rather constantly remodel and look rather delicate under the microscope. In the novel "The Secret Garden," a frail young boy with an unspecified spine problem is overly confined indoors in what seems to be a case of rickets (vitamin D deficiency). Only when encouraged to play outside in the fresh air and sun does he begin to become healthy [3]. Because of the concerns of skin cancer, some people may limit outdoor skin exposure by staying indoors or covering their bodies with sunblock or clothing. Some bicyclists develop osteoporosis due to the use of calcium in muscles and lack of impact-stimulated bone remodeling [4]. Age is also a risk factor in developing osteoporosis, which is important in our aging population. There is also evidence that having healthy vitamin D levels reduces the risk of some cancers such as colorectal cancer and pancreatic cancer [5]. Vitamin D deficiency is common in the USA, especially in minority groups. In the developed world, dairy products are often fortified with vitamin D.

Coenzyme Q10 is a key component of the electron transport chain in the mitochondria of all higher life forms, and is thus so ubiquitous in nature that it was given the name ubiquinone. Coenzyme Q10 has the vital function handing off electrons during production of the proton gradient that regenerates ATP cellular energy. Its electron grabbing ability makes it a powerful intracellular antioxidant. Ubiquinol is the reduced form of ubiquinone. Ubiquinol is able to cross cell membranes and therefore has improved bioavailability when taken as a supplement. Our bodies make coenzyme Q10 in a lengthy 15 step process. Our coenzyme Q10 levels are highest at age 20 and then steadily decline. In several countries, coenzyme Q10 is a prescribed medication intended to strengthen heart muscle in cardiac patients [6]. However, in the USA, it is relegated to the status of a dietary supplement. It does exist in many foods, but in generally low concentrations. One would have to eat massive amounts of meat or fish to get a substantial amount. Part of the coenzyme Q10 synthetic pathway is shared in the synthesis of cholesterol. Statin medications block this common (mevalonate) pathway, lowering cholesterol, but also lowering coenzyme Q10 levels. Side effects of statin drugs included liver disease and muscle weakness, possibly due to the negative effects on mitochondrial function. Since statin medications are beneficial to many people, the risk/benefit must be considered. Diet and exercise are usually recommended first, before a lifetime statin prescription. Some claim that coenzyme Q10 is the antidote to statin-induced mitochondrial toxicity, and that it should be taken along with the statin as a preventative measure [7]. Coenzyme Q10 is naturally occurring and considered very safe. It may help athletic energy production, especially as we get older. There has been no solid evidence that the poorly absorbed form of coenzyme Q10 (ubiquinone) significantly increased athletic performance in healthy athletes. However, a recent study using higher doses of the better absorbed form of coenzyme Q10 (ubiquinol) showed a 2.5 percent improvement in peak power in German Olympic athletes [8].

Acetyl-L-carnitine is naturally produced in the body and helps transport fatty acids into the mitochondria for fat burning in the Krebs cycle. The acetyl component of L-carnitine is thought to improve bioavailability. Supplementation with acetyl-L-carnitine may help fat burning energy production when combined with exercise [9]. It also has a powerful antioxidant effect, especially in protection of the phospholipids of cell membranes.

R-lipoic acid is another antioxidant that helps restore other antioxidants (the glutathione system as well as intracellular vitamin C and E). It is soluble in both water and fat, so it can reach all parts of the body, including the brain. Given to old rats, it restored declining mitochondrial membrane potentials and increased oxygen consumption [10].

Magnesium (Mg) is an essential mineral bound to every molecule of ATP (adenosine triphosphate) in the body. Mg is a required cofactor in more than 300 enzymes (proteins which drive the chemical reactions in the body) including those that create ATP energy. Mg helps muscle relaxation and may help reduce muscle cramps and reduce airway constriction in asthma. Mg citrate is the form best absorbed as a pill, but can be bulky. If one reads a multivitamin nutritional label, all of the other essential minerals are usually included at 100 percent of the recommended dietary allowance (RDA), except for Mg which is usually only supplied at 25–30 percent of what you need. Including 100 percent of the RDA for Mg would make the tablet too large. Another reason athletes may be deficient in Mg is because it is lost in sweat. Dietary Mg may be lacking in overly processed foods. Natural food sources of Mg include whole grain cereals, leafy greens, nuts, and seeds.

Resveratrol is an antioxidant found in plants, notably in the skin of red wine grapes. The French Paradox is the finding that the French have a high fat diet, yet have less heart disease, less obesity, and a longer lifespan than Americans. One possible explanation is red wine consumption, and resveratrol is thought to be the active ingredient in red wine. The French Paradox is controversial as there may be other reasons (the French eat more good fats and have a better overall diet) and the small amount of resveratrol in wine does not seem to be enough to be of benefit. However, there are several small lab creatures with short measurable life spans that live longer when given resveratrol [11]. Resveratrol works in the mitochondria in at least three ways: as an antioxidant, as an activator of mitochondrial biogenesis via PGC-1a, and by protecting against age-related mitochondrial decline [12]. Resveratrol can be taken in a concentrated pill form (without the alcohol).

Fish oil is the reason why Greenland Eskimos have a very-high-fat diet of salmon and seal meat/blubber (seals eat salmon) yet have an extremely low rate of heart disease [13]. It is not just genetic because Eskimos who eat western diets also have high western rates of heart disease. Fish oil (Omega-3 fat) is a prime example of why not all fats are bad. Fats are an essential part of our bodies, especially our cell membranes. Since mitochondria are largely composed of membranes and depend on a strong inner membrane to generate ATP, it would make sense that healthy dietary fats will help build healthy membranes and thus create healthy mitochondria. Fish oil is in oily cold water fish, but can also be taken in gel cap form if mercury is a concern. Purified fish oil is mercury free. Fish oil can lower the risk of heart disease with little or no side effects compared with modern medications. It can also stabilize heart electrical conductivity and decrease the risk of some cardiac arrhythmias. Fish oil has a mild blood thinning effect (decreasing blood clot formation) that can be beneficial in most people. However, if there is a bleeding problem or surgery is planned, the fish oil supplementation should be stopped.

Olive oil is a key component of the Mediterranean Diet, shown to be the healthiest diet in the world with lowered cardiac disease risk. Olive oil is a monounsaturated (good) fat that enriches membranes and can help lower low-density lipoprotein (LDL) (bad) cholesterol. Virgin olive oil (the first pressing) is rich in vitamins and antioxidants. It is easy to add to one's diet. A small teaspoon in a blended protein shake adds creaminess without a strong olive aftertaste.

Multivitamins with iron are a way to cover dietary deficiencies and promote healthy RBC production. The iron-based hemoglobin in our RBCs carries oxygen to our tissues. Healthy oxygen delivery maximizes the efficient aerobic energy system and also speeds active recovery from anaerobic exertion. Iron levels can be low in vegetarians and menstruating women. Most men eat more than adequate amounts of red meat. Too much iron can be harmful to the liver and heart, especially in a genetic disorder called hemochromatosis, which is treated with bloodletting. Erythropoietin (EPO) is a hormone produced by the kidneys that increases RBC production. EPO is useful in treating sick patients such as renal failure patients with anemia and cancer chemotherapy patients with anemia, as it can boost their energy levels. When EPO and/or blood doping (concentrating one's hematocrit with stored RBCs) is abused by healthy athletes, it confers an unfair advantage and risks dangerous blood clots. For this reason, EPO and blood doping are banned and severely penalized in sports.

Fiber lowers serum cholesterol by interrupting the bile acid cycle. Bile acids are made from cholesterol in the liver and released into the intestines by the gall bladder when we eat a fatty meal. The bile acids help fat digestion and absorption. The bile acids are then normally reabsorbed through the intestines and recycled. Dietary fiber binds the bile acids which are then eliminated, breaking the bile acid cycle. To make more bile acids, the liver grabs more LDL (bad) cholesterol from the blood stream, thus lowering serum LDL cholesterol. Fiber can come in the form of oatmeal, whole grains, fruit, and vegetables. The American diet of meat and processed food is woefully lacking in fiber. That is why most can benefit from supplementary dietary fiber such as psyllium husk. The more fiber in the diet, the more serum cholesterol can be lowered. In some people who add sufficient fiber to their diet, the need for cholesterol-lowering statin medications could be reduced or discontinued [14]. Very large intake of dietary fiber can interfere with the absorption of good fats (fish oil and olive oil) and fat soluble vitamins.

In summary, there are many vitamins, minerals, and nutrients that support metabolism and cell energy. However, this chapter is meant to highlight the need for a balanced and varied diet, rather than relying on a dozen supplement pill bottles. It is important to not succumb to trendy fad diets that focus on only one item or food group and thus limit the variety of the foods we eat. We have all been told that health and fitness depends on a good diet and adequate exercise. Hopefully, this chapter gives more detail on the nutritional part of the equation.

References

[1] Sebastian RS, Cleveland LE, Goldman JD, Moshfegh AJ. Older adults who use vitamin/mineral supplements differ from nonusers in nutrient intake adequacy and dietary attitudes. Journal of the American Dietetic Association 2007;107(8):1322–32.

[2] Hallberg L, Brune M, Rossander L. The role of vitamin C in iron absorption. International Journal for Vitamin and Nutrition Research - Supplement 1989;30:103–8.

[3] Burnett FH. The Secret Garden. New York: Frederick A. Stokes; 1911.

[4] Campion F, Nevill AM, Karlsson MK, et al. Bone status in professional cyclists. International Journal of Sports Medicine 2010;31(7):511–5.

[5] Garland CF, Gorham ED, Mohr SB, Garland FC. Vitamin D for cancer prevention: global perspective. Annals of Epidemiology 2009;19(7):468–83.

[6] Marcoff L, Thompson PD. The role of coenzyme Q10 in statin-associated myopathy: a systematic review. Journal of the American College of Cardiology 2007;49(23):2231–7.

[7] Zlatohlavek L, Vrablik M, Grauova B, et al. The effect of coenzyme Q10 in statin myopathy. Neuroendocrinology Letters 2012;33(Suppl. 2):98–101.

[8] Alf D, Schmidt ME, Siebrecht SC. Ubiquinol supplementation enhances peak power production in trained athletes: a double-blind, placebo controlled study. Journal of the International Society of Sports Nutrition 2013;10:24.

[9] Yano H, Oyanagi E, Kato Y, et al. L-carnitine is essential to beta-oxidation of quarried fatty acid from mitochondrial membrane by PLA(2). Molecular and Cellular Biochemistry 2010;342(1-2):95–100.

[10] Hagen TM, Liu J, Lykkesfeldt J, et al. Feeding acetyl-L-carnitine and lipoic acid to old rats significantly improves metabolic function while decreasing oxidative stress. Proceedings of the National Academy of Sciences of the United States of America 2002;99(4):1870–5.

[11] Valenzano DR, Terzibasi E, Genade T, et al. Resveratrol prolongs lifespan and retards the onset of age-related markers in a short-lived vertebrate. Current Biology 2006;16(3):296–300.

[12] Ungvari Z, Sonntag WE, de Cabo R, et al. Mitochondrial protection by resveratrol. Exercise and Sport Sciences Reviews 2011;39(3):128–32.

[13] O'Keefe JH Jr, Harris WS. From inuit to implementation: omega-3 fatty acids come of age. Mayo Clinic Proceedings 2000;75(6):607–14.

[14] Moreyra AE, Wilson AC, Koraym A. Effect of combining psyllium fiber with simvastatin in lowering cholesterol. Archives of Internal Medicine 2005;165(10):1161–6.

Your Body at Work, Play, and Rest

Building fitness

If you purchased this book, you are likely interested in learning how to improve your fitness. This might either be for health reasons, to prevent cardiovascular disease, to lose some weight, to become more competitive, to gain more energy, or to get more enjoyment from physical activities. The human body has the remarkable capacity to adapt to physical training by becoming more powerful, leaner, faster, and more energetic. Physical fitness is built deep inside our bodies. Fitness and the expectation of beneficial change is the reason why we motivate ourselves to exercise. However, it takes a lot of hard work and commitment. It used to be thought that top athletes had natural-born ability and that their amazing physical performances came easily. However, this is a myth. Athletes have to work hard (and top athletes extremely hard) to achieve high levels of performance. This is summed up succinctly in Greg's quote, "It never gets easier, you just go faster." With enough motivation and commitment you can build fitness. Hopefully, this book explains all of the wonderful benefits of exercise to help you become motivated or stay motivated. The commitment must come from you.

A major theme of this book is that fitness is built one cell at a time, by the changes that occur inside your cells and inside your mitochondria. Although these changes manifest themselves with measurable changes in performance and in visual appearance as our bodies transform, it all has to do with a biologic response to exercise at the cellular and subcellular level.

This chapter will delineate the current thinking in exercise physiology, a subject of tremendous scientific research. There are changes that occur inside your body and inside your cells when you exercise. We will integrate what we have discussed thus far, so you can understand how your body works when you train and how your body and your cells respond to exercise. We will then reveal which training methods are the most effective in increasing fitness and improving health. The beneficial changes to your body include: increased O_2 consumption, increased muscle mass, increased power, fat burning, and increased endurance. Topics covered in this chapter include: aerobic versus anaerobic exercise, the different muscle fiber types, fatigue and recovery, lactate threshold and clearance, carbohydrate metabolism, fat burning, cardiac output, skeletal muscle power, and endurance.

Cardiovascular adaptation to endurance exercise

It is easy to see why athletes focus on their skeletal muscles. Muscles make up 40–50 percent of your body mass and are the most prevalent tissue in the body (except for cases of morbid obesity). Muscles are what exert the contractile forces that move our

The Science of Fitness: Power, Performance, and Endurance. http://dx.doi.org/10.1016/B978-0-12-801023-5.00006-5

Figure 6.1 Rowing engages the muscles of the arms, shoulders, core, back, and legs with an intensity that triggers mitochondrial biogenesis.
Photography: public domain, US Government.

bodies when we exercise (Figure 6.1). Muscle output is what is measured with power metering. Muscles are what ache when you train hard. Muscles are what change visibly when you improve body composition and tone. However, some of the fittest athletes in the world are endurance athletes who do not have enormous muscles. What endurance athletes build the most is a muscle deep inside the body, the heart. In the short term, the muscle of the heart can change its output dramatically from a basal resting state to a very high maximum output during strenuous exercise. In the long term with regular exercise, the heart can make remarkable changes to adapt to the physical challenges of training.

The first marathon
In the late summer of 490 BC, the Greeks defeated the Persians at the Battle of Marathon. According to legend a Greek messenger, Pheidippides (or Philippides) was dispatched to announce the victory to the leaders in Athens and to warn of a possible Persian naval counterattack on undefended Athens. After participating in the battle himself, he ran the entire distance without stopping, and upon reaching the Athenian magistrates exclaimed, "Nikomen!" (We have won!), before collapsing and dying [1] (Figure 6.2).

The marathon race is still considered a supreme test of human endurance. But if the first marathoner died from the ordeal of running 26.2 miles, is it really a healthy

Figure 6.2 Luc-Olivier Merson's painting of Pheidippides announcing victory after the Battle of Marathon.
Photography: public domain.

endeavor? Maybe we can learn lessons from Pheidippides. Prior to the battle, he had no doubt marched great distances with his army. Although he may have specialized as a herald, he probably participated in the battle himself. Likely selected for his running ability and lack of major battle injuries, it would not be surprising if he had incurred some minor injuries. Time was of the essence and he was no doubt sent on his journey in sleep-deprived condition, with no time for recovery, and possibly dehydrated. Primitive socks were used at the time, but footwear consisted of flat sandals, and he may have run barefoot. There was no organized support for his run, whereas today marathon races have nutritional and hydration support, medical backup, and logistical support. Although it is rare, sometimes extreme duration athletic events can be damaging or even fatal. Preparation, planning, training, rest, recovery, nutrition, and hydration are essential parts of running a marathon. There are some indications that ultra-long distance athletic events can cause free radical damage [2]. The Tour de France has been described as running a marathon every day for 3 weeks. Although a testament to human endurance, is the Tour also too grueling on the human body?

The heart pumps oxygen and fuel to our body, brain, organs, and muscles. In the process, the heart uses tremendous amounts of energy, especially during strenuous exercise of any significant duration. The heart can never take a break from its work and therefore needs ample oxygen, fuel, and mitochondria to accomplish its work. The very first oxygen-containing arteries that come from the heart (to supply itself) are the coronary arteries. In arteriosclerosis, narrowing or occlusion of the coronary arteries can be a limiting factor in physical activity. Coronary artery disease can present as "exercise intolerance," as in the case of President George W. Bush. A stress test revealed abnormal EKG findings with exercise, due to a greater than 95 percent occlusion of a coronary artery, successfully treated with a stent [3]. When it comes to energy supply, the heart is fuel flexible. It can use carbohydrates (glucose and glycogen), fat (actually the primary fuel source for the heart), and broken down protein (in times of starvation). Myocardial mitochondria burn these fuels in the presence of oxygen to create the energy that empowers our heart and cardiac output. When trying to make an automotive engine run, you only need three ingredients: fuel, oxygen, and spark. Mitochondria supply the spark. Mitochondria are so important to heart energy that they occupy up to 1/3 of the myocardial tissue volume. When we exercise enough to make our hearts beat faster and more forcefully, the heart responds to the challenge by becoming a stronger muscle and by becoming a more efficient pump. With training, the heart also uses more energy by multiplying its mitochondria and thus consuming more fuel and oxygen. The trained heart can become so efficient with each beat [4], that top endurance athletes can lower their heart rates to the 50s or even the 40s (in beats per minute). By moving more oxygenated blood with every heartbeat, the well-trained endurance athlete can perform for hours and hours.

With exercise, many other organs besides the heart improve function with increased blood flow and increased metabolism. The liver becomes more efficient at shifting between metabolic states, liberating carbohydrates and fat from storage, and clearing metabolites such as lactic acid. Mitochondria convert stored body fat (adipose tissue) into energy. The blood vessels dilate and become more compliant, reducing high blood pressure. Circulation to the kidneys improves for better homeostasis and fluid control. The brain is stimulated and better oxygenated. The pancreas and other endocrine organs become better at moderating blood glucose levels, for prevention of type 2 diabetes. These organs support the exercising body and are themselves improved with exercise.

Muscle adaptation to exercise

To understand how our muscles respond to training, it is useful to know about the different muscle fiber types. Ideally, we would all be super beings and have extreme muscle power that lasts all day, but in reality there is a tradeoff and we cannot have both, due to muscle fiber specialization and inherent limitations. The skeletal muscles in your body are made up of a mixture of different muscle fiber types and in cross-section with special staining, look like a mosaic [5].

Type I (slow twitch) muscle fibers are very good at aerobic endurance and are in high percentage in postural muscles and in the muscles of long distance athletes. Type I fibers have abundant mitochondria, resist fatigue, and burn fat very efficiently.

However, they also tend to be thinner in cross-section and are therefore less powerful. When we engage in low stress tasks such as raising a fork to our mouth, we engage type I muscle fibers. These low power activities can be performed all day with no appreciable fatigue. When more force is required, type II muscle fibers are called into play.

Type IIA (intermediate) muscle fibers can use both aerobic and anaerobic energy, are more powerful than type I muscles, and can contract faster. Type IIA fibers are sometimes given cursory attention by athletes and trainers because they are not as specialized as the other two types and are the least understood, but as we will see later, they can increase with vigorous training and can give the athlete a combination of both endurance and power. Type IIA fibers can give you a competitive edge and are very important when sustained power is required.

Type IIX (fast twitch, formerly called type IIB) [6] muscle fibers have the fewest mitochondria and therefore have poor fatigue resistance. Type IIX fibers rely more on glycogen and glucose and less on fat burning. However, they hypertrophy (enlarge) more readily in response to resistance exercise and can contribute to overall lean muscle mass. Because muscles with more type IIX fibers can grow thicker and contract faster, they are very powerful.

There is much debate as to whether or not we are born with a set percentage of muscle fiber types which then predetermine what sort of athlete we can become as adults [7]. For example, the physical requirements for different positions on an NFL football team are so specific, that it is not reasonable for a lineman and wide receiver to retrain and switch duties. However, we now know that training can cause a portion of the muscle fibers to switch sides for an optimal blend of power and endurance. Of the three types of muscle fibers to have, type IIA fibers are the most versatile. They can use both fat-burning aerobic energy for endurance and glycogen-burning anaerobic energy for power and speed. Interestingly, both power lifters and couch potatoes have a higher percentage of type IIX muscle fibers [8]. In patients with paralysis, type IIX fibers become the dominant muscle fiber type [9]. It seems that this is the default muscle fiber type when aerobic energy is not required. Remember that type IIX fibers have the lowest mitochondrial density. It makes sense that with lower aerobic energy requirements, that the muscles lose some mitochondrial aerobic capacity (use it or lose it). When healthy people engage in intense resistance (weight) training, there is a modest conversion from type IIX fibers to type IIA fibers (more mitochondria generating more energy) and muscle hypertrophy (increasing strength) [10]. However, if these athletes then abruptly stop exercising for 3 months, the muscle fibers revert back to type IIX fibers and actually overshoot, resulting in more type IIX fibers compared to pre-exercise [11]. Similarly, it is possible for an endurance athlete to add interval or strength training on top of long distance training to convert some type I endurance fibers into more versatile and powerful type IIA fibers [12–14]. Although overlooked in overly simplified slow twitch versus fast twitch discussions and being the least understood of the three types listed here, versatile type IIA fibers are the most fascinating and may be the most important muscle fibers in training athletes [15].

Variety in muscle fibers

Remember that in Chapter 1 we used the wildebeest as an example of mammalian muscle evolution. Wildebeests require speed to evade lions, explosive power to leap up river banks, and endurance to stay with the herd. As mammals evolved they developed different muscle fiber types to survive a variety of conditions. Although different muscles in the body can have varying proportions of the different fiber types, they are generally a mosaic of different types to give the muscle (and the animal) versatility. The Tour de France is the ultimate test of muscle versatility. To win a sprint finish, you need strength; to win a time trial you need sustained power; to win a mountain stage you need tremendous aerobic capacity; and to last until the final stage you need endurance. Although some Olympic events require very specialized muscles, Greg LeMond (and other Tour champions) had the muscle versatility to be successful at all of these disciplines. Introducing different challenges into your training will help you develop muscle versatility.

Although we only discuss three muscle fiber types (I, IIA, and IIX), there are actually several more based on biochemistry and microscopic staining characteristics. Depending on what you read, there are currently six major fiber types in human skeletal muscles: types I, IC, IIC, IIA, IIAX, and IIX [16]. It is expected that new types will be discovered in the near future. We only present three types here for simplicity's sake, knowing that there will be a lot more to learn in the future, especially in how some muscle fibers can transition into different types in response to training. The science has already come a long way from just knowing slow twitch versus fast twitch muscles.

Mitochondrial adaptation to exercise

Mitochondria multiply in response to strenuous training. This process is called *mitochondrial biogenesis* and enables individual cells and the body as a whole to adapt to increased energy demands. Elite athletes have high mitochondrial density and sedentary people have low mitochondrial density. In fact, a major component of fitness improvement is the boosting of mitochondrial mass and efficiency. This can be quantified with VO_2 max (a measurement of oxygen delivery and oxygen consumption) and watts output (i.e., power metering). The signal and mechanism that triggers mitochondria biogenesis has recently been discovered. We will explain how this new found knowledge can be applied in the practical pursuit of increasing athletic performance.

Mitochondria create cellular energy by recycling low energy ADP (adenosine diphosphate, with two phosphates) into high energy ATP (adenosine triphosphate, with three phosphates). ATP can be used directly by muscle fibers to power muscular contraction, and in the process muscles convert the ATP back into ADP. As long as there is ample oxygen, the mitochondria can generate sufficient ATP to meet the muscle's energy demands and ATP/ADP levels reach a steady state, even during exercise. However, when more power and more energy expenditure are required, oxygen becomes scarce and muscle metabolism shifts to anaerobic conditions. During anaerobic stress, the Krebs cycle slows down and the cell reverts to far less efficient glycolysis. ATP depletes and ADP accumulates. This is a low energy condition where the mitochondria are not making enough ATP. However, muscles use a "trick" to extract more ATP under these strenuous conditions. One molecule of ADP can give a phosphate (P_i) to another ADP to form ATP (to be used as energy) and AMP (adenosine monophosphate, with one phosphate, with very low energy). At rest and at low to moderate levels of exercise, AMP levels are very low. But when strenuous exercise is performed AMP levels begin to rise. For every ATP depleted, AMP levels can increase six times [17]. At low to moderate levels of exercise, our mitochondria are very good at using oxygen to make ATP energy. But with high intensity exercise, when oxygen is consumed and we switch to anaerobic glycolysis, we create more metabolites such as lactic acid. This is what is meant by *lactate threshold*, when lactic acid begins to accumulate.

Adenosine monophosphate activated protein kinase (AMPK) is an enzyme complex that detects the AMP:ATP ratio as the signal of energy shortage [18]. When the AMPK metabolic switch is activated by high AMP levels (and low ATP levels), the body is triggered into fat burning, ketosis (turning protein into fuel), improved insulin response, and increased glucose uptake by cells. AMPK also activates peroxisome proliferator-activated receptor gamma coactivator-1alpha (PGC-1a). PGC-1a is a master-control protein (encoded by nuclear DNA) that triggers mitochondrial biogenesis (creation of new mitochondria) by activating the more than 1000 genes needed to make new mitochondria.

$ADP + P_i \rightarrow ATP$	Mitochondria make ATP
$ATP \rightarrow ADP + P_i$	Muscles use ATP to do work
$2ADP \rightarrow ATP + AMP$	Two ADP can make ATP and AMP
$AMP:ATP \rightarrow AMPK$	A high AMP:ATP ratio activates AMPK
$AMPK \rightarrow PCG\text{-}1a$	AMPK activates PCG-1a (master control)
$PCG\text{-}1a \rightarrow mitochondria$	PCG-1a triggers mitochondrial biogenesis
$Mitochondria \rightarrow ATP$	More mitochondria make more ATP

In short, AMP elevation and ATP depletion is the signal indicating that the mitochondria cannot meet current cellular energy demands. The AMP:ATP ratio is difficult for us to directly feel or measure, but we may feel the parallel effects of increased lactate. The cell nucleus responds to this state of energy depletion by signalling the mitochondria to multiply so the muscles will have more ATP energy to meet anticipated higher energy demands. Sedentary people have low mitochondrial density and tend to store calories rather than burn calories. Trained athletes have high mitochondrial density. As we age, our mitochondria tend to decrease in number. However, with frequent and strenuous exercise, it would not be surprising for an active older person to have more mitochondria and more energy than a sedentary younger person.

How can mitochondrial adaptation to exercise be applied to athletic training? The implication is that quality workouts (high effort) are more beneficial than quantity workouts (high mileage) because you can get equal or better results with far less training time [19,20]. The traditional regimen of 2 days per week of pure interval training broken up with long slow recovery training may be obsolete. Since the mitochondrial biogenesis switch is turned on with the first few intervals of a ride, adding more and more intervals after those may be of diminishing benefit and could risk injury or muscle tears. When interval training goes too long, at a certain point the pain becomes more than the gain. Long slow training sessions (on days between interval workouts) may be of little mitochondrial boosting benefit [21]. A better regimen would be to train frequently and perform at least a few high intensity efforts with each and every training session [22]. Seek out hills and attack them. Push yourself when facing strong headwinds. Train at distances short enough that you do not need to hold back. Sprint at the end and finish strong.

Muscle fatigue

Fatigue is a major limitation to physical performance. It would be great if we could go all out at 100 percent effort all day and all night, but that is not possible. Fatigue can be due to mental exhaustion or sleep deprivation. Fatigue also can be due to diseases such as chronic fatigue syndrome, myasthenia gravis, or thyroid disorders. However, this section will focus on the types of muscle fatigue commonly experienced by athletes during exercise. When a nerve impulse comes down from the brain, it triggers the release of calcium (Ca^{++}) from the sarcoplasmic reticulum of the muscle cell, which then causes the actin and myosin filaments in the muscle fibers to interact, consume ATP energy, and contract. In a fresh unfatigued muscle this occurs normally. With fatigue, the muscle no longer responds normally to the nerve signal, either due to interference with the Ca^{++} release or due to problems with the actin/myosin interaction. Fatigue seems simple at first glance; we have all experienced it. We know it can curtail a hard effort and is a limitation to performance, but we also know it goes away as the pace slows down or if we take a rest. However, muscle fatigue is actually more complex than what was originally believed for many decades and has only recently been explained. Fatigue was thought to be either lack of fuel or build-up of waste products (namely lactic acid), but the current thinking is more complex and lactic acid is no longer the primary suspect.

Three types of muscle fatigue

There are different types of muscle fatigue, experienced by different types of athletes. A power lifter may only perform one to three lifts at or near maximal weight. This builds muscle because very high resistance engages the highest number of muscle fibers and is the greatest stimulus to muscle hypertrophy (growth). If the aim is sheer muscle mass, high resistance and low repetitions (reps) are the way to do it. In the weight room you will often hear, "How much can you bench press?" and you do not hear, "How often can you bench press." By definition, you cannot perform your maximum bench press weight many, many times. At the expense of all out strength, the muscle cannot perform again immediately (due to this first type of muscle fatigue). For the cyclist, runner, or swimmer who trains with high intensity interval training (HIIT), another type of fatigue is involved. The interval can be of varying time, but is generally fairly brief because you are exercising near maximum power and cannot sustain it for very long. HIIT is the most effective way to trigger mitochondrial multiplication as described in the previous section, because it leads to the energy depleted state that turns on the mitochondrial biogenesis signal. Furthermore, with HIIT you can achieve this state many times in a single workout to turn on the switch multiple times. However, each intense interval is followed by an even longer period of recovery. You must wait for the muscle fatigue to end before exerting yourself again, but clearly this is a different mechanism of fatigue than experienced by the power lifter. For the extreme long-distance endurance athlete, fatigue has yet another meaning. By their nature, endurance athletes can perform for a very long time, but not forever. They will eventually succumb to fatigue, but from yet another fatigue mechanism.

Causes and mechanisms of muscle fatigue include:

- Fuel shortage
- Dehydration
- Muscle micro tears and muscle injury
- Muscle swelling and pain
- Free radical damage
- Build-up of metabolites

Fatigue can occur when your muscles do not get enough fuel. The nerve signal comes down and the Ca^{++} is released, but there is not enough ATP energy for actin and myosin to interact and therefore the muscle cannot contract normally. This can be prevented with training, pre-event carbohydrate loading, and mid-event refuelling (eating during the workout or race). In the most severe form, glycogen stores are completely depleted and the carbohydrate fuel tank is empty [23]. This can happen in long-distance events when the athlete does not prepare, but can also occur suddenly when warning signs of hunger are ignored. "Bonking" or "hitting the wall" is extremely uncomfortable, often painful, and sometimes nauseating which can mask the warning hunger feeling. If an athlete is unfortunate enough to experience bonking once, they will usually do whatever they can to prevent it from happening again. Glycogen depletion can also occur at moderate distances early in the season or with untrained athletes because: there are fewer mitochondria in undertrained muscles, the muscles are more dependent on the limited supply of glycogen, and there is a lower percentage of fat burning type I and type IIA muscle fibers. This can be felt early in the season when athletes extend their training from short to moderate distances. During this transition in their training, they may have strong sensations of hunger and overwhelming thoughts of food. It is best to listen to these warnings and refuel. Because it takes time to digest and absorb calories, it is a good idea to eat before you get hungry. Of course this must be balanced with not overeating. Too many calories, even when exercising regularly can lead to unwanted weight gain. There is also evidence that caloric restriction (under eating) enhances the mitochondrial biogenesis signal [24]. "Bonk training" uses the principle of caloric restriction in combination with exercise to get the most from training. However, we do not recommend it because of the risk of severe fatigue from glycogen depletion, undo discomfort from denying your hunger, the risk of self-destructive anorexic behavior, and the risk of turning your hard-earned muscle mass into fuel (ketosis). Ketosis is an evolutionary reaction to starvation that allows us to survive at the expense of turning our muscle protein into energy. Ketones (metabolites of protein breakdown) can be measured in the blood stream and the acetone produced can be detected in the breath [25].

Dehydration is another preventable cause of fatigue. If you are volume depleted, your circulatory system will not work as efficiently, your heart will work harder than it has to, and oxygen delivery to your muscles is impaired. In hot weather, there is risk of heat exhaustion and heat stroke. Drinking and carrying extra fluids when it is hot can prevent this from happening. Refer to Chapter 2 (Homeostasis section) to recognize and prevent overheating. Modern fabrics can do much to cool the body. By wicking away water, the evaporative cooling effect is enhanced and comfort is increased. Arm coolers are specially designed fabric sleeves that when are damp with sweat or splashed with water, cool more effectively than bare skin and also block harmful solar rays more effectively than sunscreen lotions.

When ultra-high resistance exercise is performed, a high percentage the muscle fibers are engaged which can lead to high muscle tension and micro tears of the muscle fibers and connective tissue. Some muscle soreness is expected with this sort of exercise as it can be the trigger for greater muscle hypertrophy. However, the pain should be diffuse, dull, and in the belly of the muscle. If the pain is localized, sharp, or at the muscle attachments, there can be injury such as muscle strains and tendonitis and you will need to give it a rest. If this pain is ignored, over use can lead to muscle tearing, muscle bleeding, and tendon rupture.

When our muscles ache, it can be from build-up of metabolites, swelling from injury and inflammation, stretching of the muscle sheath, back-up of lymphatic fluid, poor arterial inflow (peripheral vascular disease or compartment syndrome), and poor venous outflow. Muscle pain is a signal from your muscles telling you to stop or slow down. Elite competitive athletes can either ignore the pain, generate natural endorphins to dull the pain, or develop very high pain tolerance. Endorphins are substances released in response to stress that stimulate opioid receptors in the brain [26]. The resulting relief of pain and feelings of euphoria has been called the "runner's high," although there is evidence that endocannabinoids are the real cause of this sensation [27].

According to Bob Roll (a former bicycle racer and currently a cycling sports commentator), "Racing is about pain and revenge." Elite competition can indeed be very painful. Even for the amateur athlete, some muscle ache is to be expected if you are training hard enough, but should be mild and temporary. If it gets worse over time, you need to look into other more serious causes.

When our mitochondria turn fuel into ATP energy, free radicals are generated in and around our mitochondria. Fortunately our mitochondria and cells have ways to neutralize these free radicals and trained athletes have better antioxidant systems than sedentary people [28]. The sulphur-based glutathione system is one such mitochondrial antioxidant system that neutralizes free radicals [29]. Antioxidant foods work at the mitochondrial level and may be protective. However, there is evidence that ultra-long distance athletic activities can overwhelm these defenses and lead to free radical damage [30]. This is the theory behind the damage that can occur to the sensitive electrical conductive system in the hearts of extreme athletes (increasing the risk of cardiac arrhythmias) and accelerate arteriosclerosis (increasing the risk of heart attack) [31]. There is evidence that shorter but more intense workouts are actually better than long distance training for building fitness and may be healthier as well by avoiding free radical damage. Ironically, sedentary people have accelerated free radical damage too. The leading theory of metabolic syndrome is a vicious cycle of poor energy, leaking mitochondrial membranes, increased free radical damage, and further lowering of cell energy [32].

The type of muscle fatigue experience by most athletes is the build-up of metabolites. It was once thought that lactic acid was to blame. Lactic acid increases when energy metabolism switches from efficient aerobic mitochondrial energy production (oxygen is available) to inefficient anaerobic glycolysis (low oxygen conditions). When this happens at an exertion level called the "lactate threshold," glucose is turned into lactic acid. However, lactate threshold is not the barrier some though it was. Cycling time trialists for example compete at an exertional level above their lactate threshold [33]. It was once thought that lactic acidity caused fatigue and the burning sensation that accompanies fatigue. However, there is now evidence against this theory. If you infuse muscles with lactate, the muscles do not shut down. In fact, this causes an increase in blood flow and an increase in oxidative metabolism [34]. The lactate portion of lactic acid can be converted by the muscles and liver back into glucose and used as fuel. The H^+ portion of the acid is not only buffered to prevent dramatic swings in acidity, but acidity by itself does not cause muscle fatigue. For these reasons, lactic acid is no longer believed to be the main culprit of muscle fatigue. Lactic acid can still be used as a marker for the switch between metabolic states, and the more efficiently your body converts lactic acid back into glucose the faster you can recover, but now other metabolites are suspected. Chloride (Cl^-) and extracellular potassium (K^+) may be involved. However the leading candidate is inorganic phosphate (P_i). Phosphate interferes with both Ca^{++} flow in the cell and the actin/myosin interaction. The liberation of P_i during strenuous exercise, its interfering with muscle contraction, and its clearing with recovery is a leading contender to explain metabolic muscle fatigue [35].

Phosphate (P_i) is produced in these energy producing reactions:

$$ATP \rightarrow ADP + P_i + energy$$
$$ADP \rightarrow AMP + P_i + energy$$
$$CP \rightarrow Creatine + P_i + energy$$

Creatine phosphate (CP) is a limited muscle energy reserve that can convert ADP into ATP. This reserve only lasts for a few seconds and is most important for very brief but explosive efforts such as power lifting.

$$CP + ADP \rightarrow Creatine + ATP$$

Detraining and loss of fitness

Frequent strenuous exercise is the single best thing we can do to improve mitochondrial fitness. Conversely, detraining markedly impairs our fitness. Inactivity for as little as one week can cause profound loss of fitness; rebuilding that fitness can take two or three times as long. The negative effects of detraining include: reduced oxygen consumption, lowered ATP generation, loss of lean muscle mass, decreased fat burning, increased insulin resistance, decreased power, and decreased endurance. Just as mitochondria are boosted with exercise, they will also atrophy and decrease in number with inactivity. Recent studies of NASA space station astronauts have revealed the deleterious effects of micro gravity on muscles and fitness [36] (Figure 6.3), with ramifications in the dangers of a sedentary, couch-potato lifestyle here on Earth. In the micro gravity environment of the Space Station, there is not enough gravity resistance to work the muscles. In space flight, muscles atrophy (get smaller), especially

Figure 6.3 Despite daily exercise, Space Station astronauts struggle to maintain muscle mass and fitness in micro gravity.
Photography: NASA.gov.

the muscle fiber types highest in mitochondria [37]. If lack of gravity weakens our muscles, the converse is also true: gravity strengthens our muscles. You should work against gravity at every opportunity: climb those stairs, run up that hill, and bicycle up that mountain.

Mitochondria are called the power plants of the cell. They increase in number to meet energy demands, much in the same way a city builds new power plants as its population grows and demands more energy. But city power plants can be expensive and are not built unless really needed. The building of more mitochondria requires activation of more than 1000 genes across two different genomes (nuclear and mitochondrial) [38]. In both cases, power plants are built with great economy and only when necessary. With detraining, mitochondrial endurance energy is lost fairly rapidly. That is why endurance athletes cannot take extended rest without risk of losing significant performance. It will typically take much longer to rebuild that level of fitness, compared with the time taken off from training. Type IIX fibers tend to be more preserved with detraining and be rebuilt faster once training resumes. However, with aging comes decrease in hormones such as testosterone with selective loss of type II fibers as we get older [39]. Type II fibers are very important for maintaining muscle mass, increasing metabolism, and fighting body fat gain. So it is important to add resistance training to your exercise regimen as you get older. That is the best way to fight age-related muscle atrophy.

Maintaining fitness

To maintain the fitness we worked so hard to build, we must continue to exercise on a daily basis. According to the US government, children and adolescents should get 60 min or more of exercise per day and most of that time should be moderate or vigorous intensity aerobic physical activity. Adults should get at least two and a half hours (150 min) each week of moderate-intensity aerobic physical activity [40]. A typical weekend warrior athlete is exercising more than the average American, but is still not accomplishing enough. The fitness gained during an active weekend can be lost during a sedentary work week. If a weekend of exercise is interrupted by bad weather or scheduling conflicts, fitness will decline even further.

A simple goal is to be physically active each and every day. This can take the form of bicycle commuting, running before or after work, stair climbing at work, gardening, or pushing a lawn mower. Although some elite athletes have the will power to over train to the point of needing an extended period of rest, most people are not exercising enough. Sleep is good of course, but for the vast majority of people who do not exercise enough, rest can be overrated.

There should really be no off season. Many athletes are accustomed to exercising during the spring, summer, and fall, with no exercise during the winter months. This will have severe negative effects on fitness and health. It is very important to develop a winter training program. If not, the mitochondria will severely diminish and require months of spring training, just to reach previous baseline fitness.

Figure 6.4 We need to stay active during all phases of our lives.
Photography: Jim Henderson.

To maintain fitness, it is very important to avoid injury. A broken weight-bearing bone can take 6 weeks to heal. Impact from running can cause joint pain, shin splints, and plantar fasciitis. Risky "extreme" sports should be avoided. Power lifting can cause muscle tears, ligament sprains, tendon strains, and back injuries. It is very important to maintain overall health, reduce infection, maintain core strength and bone health, and avoid accidental injury of all types. Automobile accidents cause 40,000 US fatalities and countless injuries every year. Staying out of the emergency room is one of the best things we can do to stick to a fitness program.

Lifestyle change is probably the biggest reason people lose fitness. Adulthood, parenthood, and increased responsibility often lead to a more sedentary lifestyle. We must remain active throughout our lives. To serve our families, we must be as healthy as possible, and that means we must continue to exercise (Figure 6.4).

References

[1]   Herodotus [Author], Grene D. [Translator]. The History. First edition. Chicago: University of Chicago Press; 1988.
[2]   Schwartz RS, Kraus SM, Schwartz JG, et al. Increased coronary artery plaque volume among male marathon runners. Missouri Medicine 2014;111(2):85–90.

[3]   DeFrank T, George W. Bush's Close Call. National Journal (online), October 11, 2013; www.nationaljournal.com.

[4]   Baggish AL, Wood MJ. Athlete's heart and cardiovascular care of the athlete. Circulation 2011;123:2723–35.

[5]   Dubowitz V, Sewry CA, Oldfors A. Muscle Biopsy: A Practical Approach, 4th edition. New York: Elsevier; 2013.

[6]   Smerdu V, Karsch-Mizrachi I, Campione M, et al. Type IIx myosin heavy chain transcripts are expressed in type IIb fibers of human skeletal muscle. American Journal of Physiology 1994;267(6 Pt 1):1723–8.

[7]   Simoneau JA, Bouchard C. Genetic determinism of fiber type proportion in human skeletal muscle. Journal Federation of American Societies for Experimental Biology 1995;9(11):1091–109.

[8]   Tanner CJ, Barakat HA, Dohm GL, et al. Muscle fiber type is associated with obesity and weight loss. American Journal of Physiology - Endocrinology and Metabolism 2002;282(6):E1191–6.

[9]   Scelsi R. Skeletal muscle pathology after spinal cord injury: our 20 year experience and results on skeletal muscle changes in paraplegics, related to functional rehabilitation. Basic and Applied Myology 2001;11(2):75–85.

[10] Staron RS, Malicky ES, Leonardi MJ, et al. Muscle hypertrophy and fast fiber type conversions in heavy resistance-trained women. European Journal of Applied Physiology and Occupational Physiology 1990;60(1):71–9.

[11] Andersen JL, Aagaard P. Myosin heavy chain IIX overshoot in human skeletal muscle. Muscle Nerve 2000;23(7):1095–104.

[12] Vechetti IJ Jr, Aguiar AF, de Souza RW, et al. NFAT isoforms regulate muscle fiber type transition without altering CaN during aerobic training. International Journal of Sports Medicine 2013;34(10):861–7.

[13] Laursen PB, Jenkins DG. The scientific basis for high-intensity interval training: optimizing training programmes and maximizing performance in highly trained endurance athletes. Sports Medicine 2002;32(1):53–73.

[14] Aagaard P, Andersen JL. Effects of strength training on endurance capacity in top-level endurance athletes. Scandinavian Journal of Medicine and Science in Sports 2010;20(2):39–47.

[15] Aagaard P, Andersen JL, Bennekou M, et al. Effects of resistance training on endurance capacity and muscle fiber composition in young top-level cyclists. Scandinavian Journal of Medicine and Science in Sports 2011;21(6):e298–307.

[16] Staron RS, Herman JR, Schuenke MD. Misclassification of hybrid fast fibers in resistance-trained human skeletal muscle using histochemical and immunohistochemical methods. Journal of Strength & Conditioning Research 2012;26(10):2616–22.

[17] Hardie DG, Salt IP, Hawley SA, Davies SP. AMP-activated protein kinase: an ultrasensitive system for monitoring cellular energy charge. Biochemical Journal 1999;338(Pt 3): 717–22.

[18] Hardie DG. Sensing of energy and nutrients by AMP-activated protein kinase. American Journal of Clinical Nutrition 2011;93(4):891S–6S.

[19] Burgomaster KA, Howarth KR, Phillips SM, et al. Similar metabolic adaptations during exercise after low volume sprint interval and traditional endurance training in humans. Journal of Physiology 2008;586(1):151–60.

[20] Daussin FN, Zoll J, Dufour SP, et al. Effect of interval versus continuous training on cardiorespiratory and mitochondrial functions: relationship to aerobic performance improvements in sedentary subjects. American Journal of Physiology Regulatory, Integrative and Comparative Physiology 2008;295(1):R264–72.

[21] Dudley GAS Abraham WM, Terjung RL. Influence of exercise intensity and duration on biochemical adaptations in skeletal muscle. Journal of Applied Physiology Respiratory Environmental Exercise Physiology 1982;53(4):844–50.

[22] Helgerud J, Høydal K, Wang E, et al. Aerobic high-intensity intervals improve VO_2 max more than moderate training. Medicine and Science in Sports and Exercise 2007;39(4):665–71.

[23] Bourne PE, Rapoport BI. Metabolic Factors Limiting Performance in Marathon Runners. PLoS Computational Biology 2010;6(10).

[24] López-Lluch G, Hunt N, Jones B, et al. Calorie restriction induces mitochondrial biogenesis and bioenergetic efficiency. Proceedings of the National Academy of Sciences 2006;103(6):1768–73.

[25] Musa-Veloso K, Likhodii SS, Cunnane SC. Breath acetone is a reliable indicator of ketosis in adults consuming ketogenic meals. American Journal of Clinical Nutrition 2002;76(1):65–70.

[26] Simantov R, Snyder SH. Morphine-like peptides in mammalian brain: isolation, structure elucidation, and interactions with the opiate receptor. Proceedings of the National Academy of Sciences USA 1976;73(7):2515–9.

[27] Raichlen DA, Foster AD, Seillier A, et al. Exercise-induced endocannabinoid signaling is modulated by intensity. European Journal of Applied Physiology 2012;113(4):869–75.

[28] Dékány M, Nemeskéri V, Györe I, et al. Antioxidant status of interval-trained athletes in various sports. International Journal of Sports Medicine 2006;27(2):112–6.

[29] Marí M, Morales A, Colell A, et al. Mitochondrial glutathione, a key survival antioxidant. Antioxidants & Redox Signaling 2009;11(11):2685–700.

[30] Schwartz RS, Kraus SM, Schwartz JG, et al. Increased coronary artery plaque volume among male marathon runners. Missouri Medicine 2014;111(2):85–90.

[31] O'Keefe JH, Patil HR, Lavie CJ, et al. Potential Adverse Cardiovascular Effects from Excessive Endurance Exercise. Mayo Clinic Proceedings 2012;87(6):587–95.

[32] Kim J, Wei Y, Sowers JR. Role of Mitochondrial Dysfunction in Insulin Resistance. Circulation Research 2008;102:401–14.

[33] Kenefick RW, Mattern CO, Mahood NV, Quinn TJ. Physiological variables at lactate threshold under-represent cycling time-trial intensity. Journal of Sports Medicine and Physical Fitness 2002;42(4):396–402.

[34] Ahlborg G, Hagenfeldt L, Wahren J. Influence of Lactate Infusion on Glucose and FFA Metabolism in Man. Scandinavian Journal of Clinical & Laboratory Investigation 1976;36(2):193–201.

[35] Westerblad H, Allen DG, Lännergren J. Muscle fatigue: lactic acid or inorganic phosphate the major cause? Physiology 2002;17:17–21.

[36] Fitts RH, Riley DR, Widrick JJ. Physiology of a Microgravity Environment Invited Review: Microgravity and skeletal muscle. Journal of Applied Physiology 2000;89: 823–39.

[37] Fitts RH, Riley DR, Widrick JJ. Functional and structural adaptations of skeletal muscle to microgravity. Journal of Experimental Biology 2001;204(18):3201–8.

[38] Horan MP, Gemmell NJ, Wolff JN. From evolutionary bystander to master manipulator: the emerging roles for the mitochondrial genome as a modulator of nuclear gene expression. European Journal of Human Genetics 2013;21:1335–7.

[39] Lexell J. Human aging, muscle mass, and fiber type composition. Journals of Gerontology Series A Biological Sciences & Medical Sciences 1995;11(6).

[40] Leavitt, MO [Secretary HHS]. 2008 Physical Activity Guidelines for Americans. Washington, DC: Department of Health and Human Services; 2008.

The Body–Brain Connection

7

Exercise and the mind

Strengthening the body strengthens the mind. This connection between physical activity and intellectualism was recognized by different cultures in antiquity. In ancient Greece, Plato recommended physical exercise and sport as a complement to education. In old Tibet, Buddhists considered the mind and body as inseparable, advocating physical training in preparation for higher learning. Scientists now understand the biologic basis of the body–brain connection. As noted by John J. Ratey, exercise boosts the brain in many ways: increasing connections between brain cells, increasing blood flow and oxygen to the brain, stimulating many areas of the brain (not just those sections involved in motor function), and increasing neurotransmitters (the vital chemicals used in nerve cell communication) [1]. There is also evidence that exercise can delay the onset and progression of neurodegenerative disorders in the brain. With our aging population, exercise has never been more important to maintain cognitive function and independence.

Brain power

The brain is composed of the cerebrum (motor control, sensation, language, spatial ability, emotion, and reasoning), the cerebellum (coordination, balance, and conservation of movement), and the brain stem (basilar functions and autonomic control) (Figure 7.1). When it comes to intellectual abilities, the cerebral cortex is where the magic happens. Whereas a rat's cerebral cortex is smooth, the human cerebral cortex is highly convoluted with gyri and sulci, increasing the surface area of the brain and increasing the number of brain cells. The cerebral cortex is called "gray matter" because it has dense layers of nerve cells (neurons). Each neuron has many finger like extensions (dendrites) which interconnect with other neurons in a highly complex network. The lobes of the cerebrum are separated by fissures (crevices) that demarcate the specialized areas of brain function. Deeper in the brain are the longer connections between the lobes and the two sides of the brain called "white matter" due to the fat insulating the axons (the long connections extending from neurons).

Einstein's brain

After his death, Albert Einstein's brain was preserved and studied. Photographs of his brain still garner interest from neurobiologists in the hope of explaining the source of his genius [2]. Although the size and weight of Einstein's brain

(continued)

The Science of Fitness: Power, Performance, and Endurance. http://dx.doi.org/10.1016/B978-0-12-801023-5.00007-7

were within the normal range, there were anatomic and cellular differences. His right frontal lobe (the seat of higher thinking) was more convoluted with four gyri instead of the normal three. The Sylvian fissure on each side was shorter than normal, and the corpus callosum (the white matter connecting the right and left sides of the brain) was thicker than normal [3]. This meant that there was more connectedness and communication between different areas of Einstein's brain. This would explain his extensive use of thought experiments, which always had to appeal to his intuitive sense and visual reasoning. Einstein's brain was also packed with more glial cells, the non-neural brain cells which support neurons in many ways, including supplying neurons with nutrient energy [4]. Not surprisingly, Einstein's brain was more energetic.

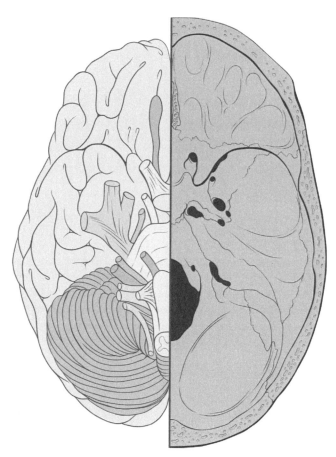

Figure 7.1 The human brain: cerebrum, cerebellum, brain stem, cranial nerves, and the skull. Medical illustration by Patrick J. Lynch.

The brain tissue is very delicate and easily torn. To prevent injury, we evolved ways to protect our brains. The scalp is loosely connected, allowing some give on impact. The skull has double layer of hard bone that forms a very effective protective shell. The brain is suspended within the skull by a bath of clear cerebrospinal fluid (CSF). Despite these built in means of protection, there are other measures to prevent brain injury, which will be described later in this chapter.

When examined at autopsy, the brain appears to be a cold lump of fat. When Leonardo Da Vinci examined human brains, he thought the brain tissue was merely support structure and that the seat of intelligence (and the soul) was in the beautifully-shaped CSF-filled ventricles inside the brain. He reasoned that the eternal soul existed without substance and therefore must inhabit the fluid cavities [5]. He was the first to employ a melted wax injection technique to preserve and sketch the shape of the ventricles. Later anatomists would prove that intelligence resided in the brain tissues. Our brains are highly metabolic: being only 2 percent of total body weight but consuming up to 20 percent of the carbohydrates. Thinking, sensing, motor control, and even basal neurologic functions require a large amount of energy that must be constantly supplied. Even a brief interruption in cerebral blood flow can have the devastating consequences of brain infarction (stroke). The selfish brain theory comes from the brain's high metabolic requirements and its ability to control the rest of the body. The brain makes sure that its nutrient needs are top priority [6]. The high metabolic activity in the brain is confirmed with modern imaging techniques. Functional magnetic resonance imaging can map metabolic activity to localized areas of the brain. Positron emission tomography scanning shows that when the body is at rest, most of sugar uptake goes to the brain and heart.

The brain is supplied by carotid arteries and vertebral arteries in the neck. Blood flow to the brain can be impaired with arteriosclerosis (hardening and narrowing of the arteries). A physician can feel the large carotid arteries in the neck to detect signs of a weak pulse or turbulent blood flow. Medical imaging can elucidate the smaller, no-palpable arteries inside the skull with magnetic resonance angiography and computed tomography angiography. As we will see later, exercise helps keep the brain well-nourished and oxygenated.

What the ancients knew

Modern science was not the first to realize the brain–body connection. Several ancient cultures recognized the inseparable nature of our minds and our physical selves. Plato established the first academy in the western world which became the prototype of the university system. Plato's curriculum included math, physics, language, biology, and physical education. Known as a great thinker, Plato was also an athlete and competitive wrestler. Besides using science to break things down to their component parts, he also was a great believer in balance especially between the body and mind. He told his students to balance intellectual studies with physical training, "to bring the two elements into tune with one another by adjusting the tension of each to the right pitch" [7].

According to legend, a 6th century Buddhist monk from India (Bodhidharma) crossed the Himalayas to Tibet with the intent of spreading his philosophy of self-enlightenment through meditation using the 49 postures of the I Chin Ching. He must have been an imposing physical entity, able to cross the Himalayan Mountains on foot and with portraits of him heavily bearded and almost barbaric in appearance. He found that his students were too frail to withstand his mental regimen. By introducing rigorous physical exercises, he strengthened both the minds and bodies of his monks. He is credited with founding the martial arts system of the Shaolin Temple, whose practitioners are capable of nearly super-human feats of strength and agility (Figure 7.2).

The Zen religion in Japan is based on the Zen arts which although are physical, are deeply rooted in mental focus, mental fluidity, and inward thinking. What the various Zen arts have in common is deep breathing as a source of power, and controlling the

Figure 7.2 Modern day Shaolin monks performing acrobatics.

flow of one's inner energy (Ki). From this combination of body and mind came the warrior spirit of physical mastery, self-control, and loyalty.

A child's mind

The brain is plastic. Not in the sense of being made of Tupperware, but in the sense of being capable of change and adaptation. Neuroplasticity is the ability of our brains to make new connections, store memories, and invent new ideas. Neuroplasticity is at its highest in childhood when we are learning and making sense of our world. Children easily learn their native language. As we get older, it becomes increasingly difficult to learn a new language. In parallel with mental development, children develop physically, increasing strength and coordination. During this crucial period of mental and physical growth, children need access to stimulating educational challenges, professional instruction, a nutritious diet, and physical exercise. What they gain neurologically in childhood will benefit them their entire adult lives. The brain learns by making connections. Each neuron in the brain has many projections called dendrites which reach out and connect to other neurons. A 2-year old has more available connections than an adult. As a child learns, these connections are retained, increased, and strengthened. Without learning these connections are lost. Just as with fitness, training the brain is a case of use it or lose it.

Besides the classroom setting of reading, writing, and arithmetics, much learning can be done outside of the classroom. Most of a child's time is spent outside of class and how that time is used greatly affects brain development. When a child watches television, images flash by stimulating the visual cortex, but they may not register intellectually because no interaction is required. In the old days, children were encouraged to play outdoors with the only instructions: do not kill yourselves and be home in time for dinner. When children explore the outdoors and learn to play amongst each other, the discovery and social interactive parts of their brains are stimulated. The strongest stimulus to learn is self-discovery through interaction, not rote memorization [8]. Nowadays, because of the fear of child predators and violence (greatly exaggerated due to news media), children are more likely to be inside, in front of a TV or preoccupied with other electronic gadgets. Television replaces creativity and imagination, discourages reading, discourages exercise, and can promote over-eating, materialism, and aggression. Excessive television viewing can create changes in the brain that lower a child's verbal intelligence quotient [9].

With attention deficit hyperactivity disorder (ADHD), children are inattentive, unfocused, easily distracted, disruptive in class, and fidgety. Exercise helps some patients with ADHD by increasing neurotransmitters in the brain (the means by which neurons communicate with each other), increasing mental focus, damping excitability regions in the brain, reducing impulsivity, and improving cerebellar motor control [10].

When school budgets are reduced, often physical education is the first target of the curriculum to be cut, with the assumption that rote memorization in the classroom is the best way to score high on standardized tests. However, exercise and play are important for childhood development in promoting physical conditioning, social interaction, emotional development, and cognitive function [11].

Changes in the brain with exercise

When we exercise, the brain benefits in many ways. With physical activity, and especially with aerobic exercise, new neuronal connections are made, and not just in the areas of motor control. The hippocampus is a more primitive part of the brain involved in learning and memorization. Exercise increases activity and development in the hippocampus, which has been confirmed with medical imaging [12,13]. Mental stress causes the release of cortisol which can reduce the volume of the hippocampus, interfere with neurotransmitters, and impair memory. Exercise helps alleviate mental and physical stress. Exercise increases the amount of beneficial neurotransmitters in the brain including: dopamine, serotonin, and glutamate. Exercise improves total body metabolism and insulin control, allowing for a steadier supply of fuel to the brain. Exercise strengthens our hearts and oxygen delivery systems, increasing brain oxygenation and reducing the risk of arteriosclerosis. Exercise allows us to better adapt to our physical surroundings and thus promotes mental flexibility and neuroplasticity.

When to daydream

When you are driving a car you are in control of a two ton machine hurling down the road at a great rate of speed. You are responsible for the safety of yourself, your passengers, and anyone else sharing or crossing the road. This is not the time to daydream. You need all of your focus on driving, which means no daydreaming, no distractions, and no text messaging. But daydreaming itself is good for the imagination and the creation of new ideas. The idea to write this book and most of its main concepts came about while I was riding my bike. When I am pedaling my bike I am traveling at a human pace, oxygen is pumping to my brain, I'm no great threat to those around me, and there are no electronic distractions. I am therefore free to daydream and ponder. When researching and writing about exercise, it was only natural to think about exercise while riding my bike.

Mark Hom

Brain injuries and prevention

As noted in the anatomy section, the brain is a soft and delicate organ. Although evolution has resulted in a skull to shield the brain from impact and cerebrospinal fluid to cushion the brain, we are still vulnerable to brain injury in our activities, sports, and lifestyles. Motor vehicle accidents account for 50,000 to 30,000 deaths per year in the US and are the leading cause of death in young people. If a virus or a war was causing 40,000 deaths per year, there would be outrage, but in a society dependent on convenient automotive transportation, this is merely the cost of doing business. A state trooper will tell you that speed kills. Actually it is not the speed that kills you, it is the sudden stopping. In a car accident, if your head strikes a firm object, your skull comes to an abrupt halt. However, your brain will keep moving and strike the inside of your skull, sometimes rebounding and injuring both ends of your brain (coup and contrecoup injury). The junction between the gray and white matter of the brain is especially vulnerable to the resulting impact and shear injury, resulting in axonal injury, concussive symptoms, and brain contusion. In addition, the arteries and veins between the skull and brain can be ripped, causing bleeding around the brain.

Signs and symptoms of a concussion may include:

- Headache
- Temporary loss of consciousness
- Confusion or mental fog
- Amnesia surrounding the traumatic event
- Dizziness or "seeing stars"
- Ringing in the ears
- Nausea
- Vomiting
- Slurred speech
- Delayed response to questions
- Appearing dazed
- Fatigue

Although high-speed deceleration injuries are especially injurious to the brain, impact injuries from sports can also cause brain damage. Although it can be entertaining to watch a solid tackle in American football, or a header goal in soccer, these activities are well known to cause concussive injures [14]. Improved headgear in contact sports and bicycling, and recent usage of protective headgear in soccer and baseball (pitchers) may lessen the damage, but the risks will still be present. Boxing (repeated blows to the head) and mixed martial arts (bare knuckle strikes, kicks, and chokes) demand high levels of physical fitness, but are too risky in terms of brain damage for us to recommend as healthy sports activities.

Cerebral infarction (stroke) is a major cause of disability and suffering but can be largely prevented. Exercise, cholesterol control, blood pressure control, blood clotting control, and smoking cessation are all effective preventative measures. Physical exam and ultrasound screening can detect arteriosclerosis sooner and there are new effective medical and surgical treatments to prevent stroke.

Why we sleep

It had long been a mystery as to why humans require sleep. There are some creatures that never sleep such as sharks that must move forward at all times (the ram effect moves water into their mouths and across their gills). Wouldn't we be more productive if we never had to sleep? It was rumored that the great inventor Thomas Alva Edison never slept (he considered it a waste of time). He did sleep, although only 3–4 hours per day with short naps, often in his laboratory. He may have invented the power nap. But most of us require 7–8 hours of sleep. Why is that? In evolutionary terms, it makes sense that we would be dormant at night, safely curled up in a tree or huddled in a cave, rather than bumping into predators or falling off cliffs in the darkness.

The mechanism of sleep has to do with the uniqueness of brain metabolism. The brain relies chiefly on blood sugar and does not burn fat. It was known that the brain also contained a limited amount of glycogen (polymerized glucose), but far less than what is in our muscles or liver. It turns out that the brain burns through this limited supply of glycogen during waking hours. Sleep allows us to rebuild these glycogen stores, to literally recharge our brains with energy [15]. Without adequate sleep, we are running our brains at low power. Chronic sleep deprivation can be a result of our modern society that demands long work hours, causes stress and anxiety, and encourages coffee and stimulants to suppress the urge to sleep in order to get the job done. However, chronic sleep deprivation impairs learning, memory, concentration, and cognitive performance. You are more apt to make mistakes, get injured, or have motor vehicle accidents. In the past, interns were indoctrinated into the medical field with long hours and high stress. Being caught sleeping was grounds for dismissal. However, the resulting sleep deprivation was the root cause of many medical errors. Currently, intern and residency hours have strict limitations and sleeping in hospital call rooms is encouraged. For athletes, sleep deprivation means decreased power and endurance, less focus, poor decision making, and slower reaction time [16].

Slowing brain atrophy

It was once thought that after childhood, the neurons in the brain stop growing and that adults should expect a steady decline in the brain function as they get older. We now know that one area of the brain is capable of growing new neurons (neurogenesis). This area is the hippocampus, a more primitive and central part of the brain associated with learning and memory. Unfortunately, the majority of the brain including the cerebral cortex stops growing, and worse yet loses cells as we age. Age-related brain atrophy (shrinkage) is a natural consequence of aging and even Albert Einstein had some. However, we can do things to slow this process. One way is to stay engaged in social and mental activities as we age. Interacting with friends and relatives is much healthier than a hermit lifestyle [17]. Mental acuity can be maintained with games and puzzles such as crosswords, learning a new hobby, artistic expression, or learning a new language [18]. We are capable of learning all throughout our lives. Exercise is another way to slow age-related brain atrophy [19], with possible mechanisms including: stimulating the hippocampus, increasing cerebral blood flow, reducing high blood pressure, and improving insulin response.

A health body and mind

In summary, there is a strong connection between the body and the mind, recognized in antiquity and confirmed by modern scientific discoveries. Both the body and brain evolved together as one and not as separate entities. The exercise we do to improve our physical bodies also helps our brains. Our brains can learn at any age as long as we stay active, both mentally and physically. Exercise increases blood flow to the brain, improves the brain's chemistry and connections, improves learning and memory, and improves brain metabolism. As the US and world population ages, exercise is one thing we can do to maintain cognitive function, independence, and self-reliance as we get older.

References

[1] Ratey JJ, Hageman E. Spark: The Revolutionary New Science of Exercise and the Brain. New York: Little Brown and Company; 2008.

[2] Falk D, Lepore FE, Noe A. The cerebral cortex of Albert Einstein: a description and preliminary analysis of unpublished photographs. Brain 2012;136(4):1304–27.

[3] Men W, Falk D, Sun T, et al. The corpus callosum of Albert Einstein's brain: another clue to his high intelligence? Brain 2014;137(4):e268.

[4] Diamond MC, Scheibel AB, Murphy GM Jr, Harvey T. On the brain of a scientist: Albert Einstein. Experimental Neurology 1985;88(1):198–204.

[5] Paluzzi A, Belli A, Bain P, Viva L. Brain 'imaging' in the Renaissance. Journal of the Royal Society of Medicine 2007;100(12):540–3.

[6] Fehm HL, Kern W, Peters A. The selfish brain: competition for energy resources. Progress in Brain Research 2006;153:129–40.

[7] Plato. The Republic. Translated by Jowett, B., Mineola, NY: Dover Publications; 2000.

[8] Dietze B, Kashin D. Playing and Learning. New Jersey: Pearson Prentice Hall; 2011.

[9] Takeuchi H, Taki Y, Hashizume H, et al. The impact of television viewing on brain structures: cross-sectional and longitudinal analyses. Cerebral Cortex 2013; Epub online.

[10] Ratey JJ, Hageman E. Spark: The Revolutionary New Science of Exercise and the Brain. New York: Little Brown and Company; 2008.

[11] Ginsburg KR. The importance of play in promoting healthy child development and maintaining strong parent-child bonds. Pediatrics 2007;119(1):182–91.

[12] Cotman CW, Berchtold NC. Exercise: a behavioral intervention to enhance brain health and plasticity. Trends in Neurosciences 2002;25(6):295–301.

[13] Erickson KI, Leckie RL, Weinstein AM. Physical activity, fitness, and gray matter volume. Neurobiology of Aging 2014;35(S2):S20–S28.

[14] Gessel LM, Fields SK, Collins CL, et al. Concussions among United States High School and Collegiate Athletes. Journal of Athletic Training 2007;42(4):495–503.

[15] Benington JH, Heller HC. Restoration of brain energy metabolism as the function of sleep. Progress in Neurobiology 1995;45:347–60.

[16] Mah CD, Mah KE, Kezirian EJ, Dement WC. The effects of sleep extension on the athletic performance of collegiate basketball players. Sleep 2011;34(7):943–50.

[17] James BD, Wilson RS, Barnes LL, Bennett DA. Late-life social activity and cognitive decline in old age. Journal of the International Neuropsychological Society 2011;17(6): 998–1005.

[18] Woodard JL, Sugarman MA, Nielson KA, et al. Lifestyle and genetic contributions to cognitive decline and hippocampal structure and function in healthy aging. Current Alzheimer Research 2012;9(4):436–46.

[19] Yuki A, Lee S, Kim H, et al. Relationship between physical activity and brain atrophy progression. Medicine and Science in Sports and Exercise 2012;44(12):2362–8.

When Things Go Wrong

Mitochondria can be negatively affected by inactivity, toxins, and genetic diseases. Many diseases of inactivity (metabolic syndrome, obesity, and type 2 diabetes) can be prevented by strengthening our mitochondria with exercise. Many inherited mitochondrial disorders have been identified. Mitochondria are suspected of having a role in degenerative diseases of the brain. Mitochondria are important in understanding cancer because they are at the center of apoptosis (programmed cell death) and tumor cells have altered metabolism, skewed toward growth rather than energy production.

Mitochondria toxins

Mitochondria are complicated biological entities and are very sensitive to a multitude of chemical toxins. Mitochondrial toxins can be divided into environmental toxins and medication-related toxins and are detailed here so the athlete can best avoid them. Mitochondrial toxicity cut short Greg LeMond's racing career. Mitochondria toxins include: cyanide, arsenic and other heavy metals, ozone, cigarette smoke, household and garden chemicals (herbicides, insecticides, and fungicides), trans fatty acids, statin medications, certain antibiotics, and some anti-HIV medications.

Cyanide (carbon triple bonded to nitrogen) specifically blocks cytochrome C in the electron transport chain of mitochondria. Cyanide is an infamous method of suicide for failed dictators and captured spies, since a small pill of hydrogen cyanide can kill quickly. No athlete of sound mind would take cyanide, but it does demonstrate the vital importance of mitochondria in life function. Low-dose cyanide exposure can occur in photography, metal processing, mining, electroplating, fumigation, and inhalation of burning plastic or rubber.

Arsenic (As) is a heavy metal that can kill by impairing mitochondria membrane potential. Arsenic is insidious because low levels are allowed in drinking water. Until recently arsenic was used to prevent wood rot due to its strong antibacterial and antifungal properties. Arsenic can leach out from treated lumber purchased in 2003 or earlier. Other toxic heavy metals include: lead (Pb), cadmium (Cd), and mercury (Hg).

Ozone (O_3) is a reactive oxidant. High in the stratosphere, ozone reacts to light and absorbs harmful UV rays. It can also be produced low in the atmosphere where we breathe. On hot, sunny, stagnant days, automobile and industrial emissions react with light and produce smog and ozone. Ozone impairs mitochondria. Visible smoke from emissions can carry "hitchhiker chemicals" that are toxic to mitochondria and your lungs. On high alert ozone days, strenuous exercise should be avoided, especially for people with pre-existing lung conditions.

The Science of Fitness: Power, Performance, and Endurance. http://dx.doi.org/10.1016/B978-0-12-801023-5.00008-9

Cigarette smoke impairs mitochondrial energy function and may disturb apoptosis, leading to lung cell necrosis. This is a suspected mechanism for chronic obstructive pulmonary disease and emphysema. Cigarette smoke is a carcinogen and the leading cause of lung cancer.

Herbicides, insecticides, fungicides, and other dangerous chemicals can be toxic to mitochondria [1]. Exposure to these chemicals should be limited and protection must be worn when handling them around the house or yard.

Trans fatty acids are partially hydrogenated vegetable oils found in processed baked goods and deep fryers. The words "partially," "hydrogenated," and "vegetable oil" do not sound ominous, but what the phrase really means is "artificial fat." Chemically altered to prolong shelf life and improve food texture, trans fats are foreign to our physiology. Dietary trans fats accumulate in our cell membranes where they alter the membrane integrity of our cells and mitochondria. Food manufacturers can include just under 0.5 g of trans fat per serving and round down to claim 0 g trans fat on the nutritional label. However, no level of trans fat is safe. If the phrase "partially hydrogenated vegetable oil" appears on the ingredient list, do not consume the product. Trans fats lower good high-density lipoprotein (HDL) cholesterol, raise bad low-density lipoprotein (LDL) cholesterol, and increase the risk of cardiovascular disease [2]. Biopsies of women's body fat have shown a correlation between trans fat content and the risk of breast cancer [3].

Medicine-related mitochondrial toxins include statins (cholesterol lowering), antibiotics, and anti-HIV drugs.

Statin medications are widely prescribed to lower bad LDL cholesterol and are effective in lowering heart disease risk in many people. However, known side effects include liver disease and muscle weakness. Statin medications block the mevalonate pathway, early in the biosynthesis of cholesterol. This is a shared pathway that also is necessary for the biosynthesis of coenzyme Q10. The vital importance of coenzyme Q10 in the electron transport chain has already been mentioned. In weighing the risk/benefit ratio of lowered cholesterol versus lowered coenzyme Q10, one must be certain that everything else is being done to lower cholesterol (such as diet, exercise, fiber, and avoiding trans fats). Coenzyme Q10 supplementation may help some people who must take statin medication. Some consider coenzyme Q10 to be the antidote to statin toxicity [4]. Coenzyme Q10 will not negate the cholesterol-lowering benefits of the statin.

Antibiotics impair or kill bacteria through different mechanisms. Since mitochondria are ancient bacteria, it is not surprising that many antibiotics are toxic to mitochondria. Linezolid is a newer antibiotic which is one of the only oral antibiotics effective against antibiotic-resistant Staph Aureus infections by inhibiting bacterial protein synthesis. It also inhibits mitochondrial protein synthesis. Long-term use of this drug can cause severe muscle weakness and lactic acidosis. Other potentially mitotoxic antibiotics include: tetracycline, minocycline, chloramphenicol, and aminoglycosides. The best way to avoid the need for antibiotics is to maintain a healthy immune system and avoid exposure. Small cuts and scrapes must be cleaned and bandaged immediately. Probiotic food such as active-culture yogurt can encourage healthy intestinal bacteria. Adequate rest and vitamin C also help the immune system. Bacterial antibiotics should not be used indiscriminately or unnecessarily (such as with viral infections where they have no benefit).

Anti-HIV drugs such as azidothymidine (AZT) have made AIDS a chronic, rather than a fatal condition for many people. Long-term treatment has revealed that the same mechanism of AZT's action (as a transcriptase inhibitor) can alter mitochondrial function by inhibiting the mitochondrial DNA polymerase gamma (the enzyme responsible for the replication of mtDNA) and thus impair mitochondrial biogenesis, resulting in muscle weakness, lactic acidosis, and lipodystrophy (a disorder characterized by accumulation of visceral fat, breast adiposity, cervical fat-pads, hyperlipidemia, insulin resistance, and fat wasting in face and limbs) [5].

Diseases of inactivity

The diseases of inactivity can occur when we do not or cannot exercise. As noted previously, we evolved as physical beings and are healthiest when physically active. When we become inactive, we lose muscle, gain fat, and create a vicious cycle of lowered metabolism and increased oxidant stress. The diseases of inactivity include:

- Obesity
- Metabolic syndrome
- Type 2 diabetes
- Failure to thrive (in adults)

Obesity is a disease. After years of saying that obesity was merely a lifestyle choice, the American Medical Association just recently declared that obesity is a disease after all. Tasked with paying for the costs of disease, the Affordable Care Act still categorizes obesity as a behavioral risk factor.

> *"Obesity as a Disease...the AMA adopted policy that recognizes obesity as a disease requiring a range of medical interventions to advance obesity treatment and prevention. Recognizing obesity as a disease will help change the way the medical community tackles this complex issue that affects approximately one in three Americans. The AMA is committed to improving health outcomes and is working to reduce the incidence of cardiovascular disease and type 2 diabetes, which are often linked to obesity."* [6]
> —The American Medical Association (AMA)

> *"(B)eing overweight or obese" is not classified as a disease but as a "behavioral risk factor"* [7]
> —U.S. Department of Health and Human Services, Patient Protection and Affordable Care Act (ObamaCare).

Semantics aside, obesity has become a worldwide epidemic affecting adults, children, developed nations, and the poor. Obesity is screened with BMI measurement. However, better predictors of health and longevity are physical fitness, abdominal girth, and body composition (fat vs. muscle).

Metabolic syndrome is a constellation of findings related to inactivity and obesity and is a major risk for developing type 2 diabetes, heart disease, and stroke. These are the criteria [8]:

- Obesity: a waist circumference of 40 in. (102 cm) or more for men and 35 in. (89 cm) or more for women.
- Hypertension: a blood pressure of 130/85 mm Hg or more.
- High blood sugar: a fasting blood glucose of 100 mg/dL or more.
- High blood lipids: blood triglycerides of 150 mg/dL or more.
- Low HDL: HDL("good" cholesterol) of less than 40 mg/dL for men, or less than 50 mg/dL for women.

Diabetes: Type 1 diabetes is often diagnosed in childhood and is when the pancreas does not make enough insulin. Type 2 diabetes (associated with obesity and metabolic syndrome) is when the body does not respond as well to insulin. In both cases, blood glucose is too high, which can result in accelerated cardiovascular disease, risk of infection, poor wound healing, neuropathy, vision problems, and organ damage.

Failure to thrive (in chronically ill adults and the elderly) is seen in nursing homes, chronic care facilities, and intensive care units where patients are confined to chronic bed rest with muscle atrophy, poor nutrition, weight loss, risk for bed sores, mental depression, and poor medical progress.

Cardiovascular disease

Cardiovascular disease is the leading cause of death and a major cause of disability with several modifiable (behavioral) risk factors including [9]:

- Hypertension (high blood pressure)
- Tobacco use
- Raised blood glucose (diabetes)
- Physical inactivity
- Unhealthy diet
- Cholesterol/lipids
- Overweight and obesity

The complications of cardiovascular disease include [10]:

- *Heart failure*: when the heart weakens and can't pump enough blood to meet the body's needs. Heart failure can result from many forms of heart disease, including congenital heart defects, cardiovascular disease (arteriosclerosis), valvular heart disease, heart infections, and cardiomyopathy.
- *Heart attack*: interrupted blood flow to the heart can damage or destroy a part of the heart muscle.
- *Stroke*: arteries to the brain are narrowed or blocked and too little blood reaches the brain. A stroke is a medical emergency—brain tissue begins to die within just a few minutes of a stroke.
- *Aneurysm*: a bulge in the wall of an artery. If an aneurysm bursts, there can be life-threatening internal bleeding. If a blood clot within an aneurysm dislodges, it may block an artery at another point downstream.

- *Peripheral artery disease*: the extremities (usually the legs) do not receive enough blood flow. It starts as leg pain when walking (claudication), but can result in severe constant rest pain or amputation.
- *Sudden cardiac arrest*: the acute, unexpected loss of heart function, breathing, and consciousness. Sudden cardiac arrest usually results from an electrical arrhythmia in the heart that disrupts its pumping action. Sudden cardiac arrest is a medical emergency. If not treated immediately, it is fatal, resulting in sudden cardiac death.

Neurodegenerative diseases

Alzheimer's disease, Parkinson's disease, and amyotrophic lateral sclerosis are the most common adult neurodegenerative diseases. Some forms of these diseases are inherited, and genes causing these diseases have been identified. As noted earlier, the brain is a highly metabolic organ that relies heavily on mitochondria energy. Studies have revealed that mitochondria could have roles in neurodegenerative diseases [11].

- *Alzheimer's disease*: alterations in enzymes involved in oxidative phosphorylation, oxidative damage, and mitochondrial binding of Aβ and amyloid precursor protein have been reported.
- *Parkinson's disease*: mutations in mitochondrial proteins have been identified and mitochondrial DNA mutations have been found in neurons in the substantia nigra.
- *Amyotrophic lateral sclerosis (Lou Gehrig's disease)*: changes occur in mitochondrial respiratory chain enzymes and mitochondrial programmed cell death proteins.

Cancer

Obesity increases the risk of several types of cancer with the following proposed mechanisms [12]:

- Fat tissue produces excess amounts of estrogen, high levels of which have been associated with the risk of breast, endometrial, and some other cancers.
- Obese people often have increased levels of insulin and insulin-like growth factor-1 in their blood, which may promote the development of certain tumors.
- Fat cells produce hormones, called adipokines that may stimulate or inhibit cell growth. For example, leptin, which is more abundant in obese people, seems to promote cell proliferation.
- Fat cells may also have direct and indirect effects on other tumor growth regulators, including mammalian target of rapamycin (mTOR) and AMP-activated protein kinase.
- Obese people often have chronic low-level inflammation, which has been associated with increased cancer risk.

Cancers associated with obesity:

- Esophageal cancer
- Pancreatic cancer
- Colon and rectal cancer
- Breast cancer (postmenopausal)
- Endometrial/Uterine cancer
- Renal cancer
- Thyroid cancer
- Gallbladder cancer

Cancer cells have an altered metabolism that suppresses efficient and aerobic mitochondrial energy in favor of more primitive and less efficient anaerobic glycolysis. It was once thought that this occurs because cancer cells outgrow their blood supply and adapt to low oxygen conditions. However, the same phenomenon occurs when there is adequate oxygen. The reason is the competition for the fuel that feeds the Krebs cycle inside mitochondria. Acetyl-coA, citrate, and glucose are diverted from the Krebs cycle to build new cells: lipids for cell membranes, nucleic acids for DNA, and amino acids for proteins [13]. Instead of burning fuel, the cell is programmed to build biomass for rapid growth. Uncontrolled cell growth is the main characteristic of cancers and this growth is fed by diverting fuel from mitochondria. The principle of tumor imaging with positron emission tomography is based on cancer's affinity for sugar. It was previously believed that this was because cancer cells merely had high metabolism, but it is really because cancer cells are in rapid building mode.

The cells in our body are constantly changing. Some cells multiply to replace others and some cells die off when they are no longer needed. Cell death is a normal and necessary process. Apoptosis is programmed cell death and determines whether a cell will continue to live or die. If it were not for apoptosis, we would have webbed fingers and toes, but instead the skin between our digits regresses during development in an orderly way. Mitochondria are central to apoptosis, responding to protein signals in the cell, releasing apoptotic proteins, and triggering the death of the cell [14]. Apoptosis research may lead to new discoveries about cancer, embryonic development, and tissue atrophy.

Mitochondrial diseases

There are many genetic mitochondrial diseases. Because mitochondria have their own mtDNA separate from the cell's nuclear DNA, some mitochondrial genetic diseases are inherited maternally: from mother to daughter, and from mother to son. Over the eons, many mitochondrial genes migrated to the nucleus, so some mitochondrial genetic diseases are inherited from both parents via classic Mendelian inheritance patterns. Mitochondria are in the cells of all of our organs and tissues, so different mitochondrial diseases can manifest in a multitude of ways. Mitochondrial diseases are difficult to diagnose and often require enzyme analysis from living tissue (muscle biopsy). Referral to an appropriate research center with mitochondrial specialists and specialized testing equipment is necessary. With physicians experienced in diagnosing mitochondrial diseases, diagnoses can be made through a combination of clinical observations, laboratory evaluation, brain imaging, and muscle biopsies. Despite these advances, many cases do not receive a specific diagnosis.

Mitochondrial diseases (from the United Mitochondrial Disease Foundation) [15]:
Alpers Disease
Barth Syndrome, LIC (Lethal Infantile Cardiomyopathy)
Beta oxidation Defects
Carnitine-Acyl-Carnitine Deficiency
Carnitine Deficiency
Creatine Deficiency Syndromes
Coenzyme Q10 Deficiency
Complex I Deficiency
Complex II Deficiency
Complex III Deficiency
Complex IV Deficiency, COX Deficiency
Complex V Deficiency
CPEO (Chronic Progressive External Ophthalmoplegia syndrome)
CPT I Deficiency (Carnitine Palmitoyltransferase I deficiency)
CPT II Deficiency (Carnitine Palmitoyltransferase II deficiency)
KSS (Kearns-Sayre Syndrome)
Lactic Acidosis
LBSL (Leukoencephalopathy, brain stem and spinal cord involvement, lactate elevation)
LCAD (Long-Chain Acyl-CoA Dehydrogenase deficiency)
LCHAD (Long-Chain 3-Hydroxyacyl-CoA Dehydrogenase deficiency)
Leigh Disease
Luft Disease
MAD (Multiple Acyl-CoA Dehydrogenase deficiency)
MCAD (Medium-Chain Acyl-CoA Dehydrogenase deficiency)
MELAS (Mitochondrial Encephalomyopathy Lactic Acidosis and Stroke-like episodes)
MERRF (Myoclonic Epilepsy and Ragged-Red Fiber disease)
MIRAS (Mitochondrial Recessive Ataxia Syndrome)
Mitochondrial Cytopathy
Mitochondrial DNA Depletion
Mitochondrial Encephalopathy
Mitochondrial Myopathy
MNGIE (Myoneurogastrointestinal Disorder and Encephalopathy)
NARP (Neuropathy, Ataxia, and Retinitis Pigmentosa)
Pearson Syndrome
Pyruvate Carboxylase Deficiency
Pyruvate Dehydrogenase Deficiency
POLG Mutations (DNA polymerase gamma mutations)
Respiratory Chain
SCAD (Short-Chain Acyl-CoA Dehydrogenase deficiency)
SCHAD (Short-Chain 3-HydroxyAcyl-CoA Dehydrogenase deficiency)
VLCAD (Very Long-Chain Acyl-CoA Dehydrogenase deficiency)

Greg LeMond's mitochondrial myopathy

The mitochondrial diseases listed above are identified by specific enzymatic deficiencies. Greg LeMond was not born with any mitochondrial deficiencies, in fact he had very strong mitochondria as evidenced by his superior athletic ability. Unlike the genetic diseases listed above, Greg LeMond contracted an acquired form of mitochondrial myopathy.

After winning the Tour de France in 1986, Greg began the next season racing and training in Europe when he crashed and broke his left hand. He felt fortunate because he could still fly fish with his right arm. While recuperating in the U.S. in the spring of 1987, Greg went on an impromptu turkey hunting expedition in a remote area of California with his uncle and brother-in-law. Turkeys have keen vision, so hunters must lie in wait from concealed positions. In a hunting accident, Greg's back and right side were blasted with lead #2 shot pellets each 0.15" (3.8 mm) from a 12 gauge shotgun. The damage was extensive with collapse of his right lung, 5 pellets lodged in or around his heart, 5 pellets lodged in his liver, multiple pellets in his lungs, some pellets ripping through and through, an estimated 60 percent loss of his blood volume, and perhaps being only 15 minutes from death. If not for helicopter evacuation to University of California Davis Medical Center, he would have died out in the field. After surgery, three dozen lead pellets remained in his body and in vital organs. Removing all of them would have been too dangerous.

The severe injuries and painful recovery resulted in Greg losing a tremendous amount of muscle mass and fitness. In the first race he entered after the accident, he could only ride 1 mile. In the second race, he could only ride 12 miles. Despite an intestinal complication requiring a second operation, teams giving up on him, and a severe fitness deficit, Greg eventually came back to the sport he loved. With a disappointing start in the 1989 Giro d'Italia, Greg was ready to quit racing altogether, but with the encouragement of his wife Kathy, he tried to endure for a few more days. Toward the end of the Giro, he started feeling stronger and in the final Giro time trial he rode as hard as he could to test himself, beating his arch rival, Laurent Fignon. Greg knew he could be a Tour contender again. LeMond went on to win the 1989 Tour de France later that season by the closest margin ever (8 seconds), the 1989 world championship to cap his amazing comeback year, and then the 1990 Tour de France for his third and last Tour de France victory.

In subsequent years, Greg's performance began to falter. The same riders he could easily drop on the mountain climbs were now dropping him. In an agonizing and frustrating period of his career, his body was letting him down. He could not figure out if he was overtraining or undertraining. Nothing was working. As it turned out, the shotgun pellets deep inside his body were leaching elemental lead into his blood stream and poisoning his mitochondria. Lead competes with calcium, binding to and interfering with critical mitochondrial enzymes. The result was altered cell calcium flow, impaired electron transport, increased reactive oxygen species, impaired antioxidant defenses, disrupted membrane ion channels, and ultimately diminished ATP energy generation from his mitochondria. Normally a single bullet does not cause lead poisoning because scar tissue surrounds the bullet and minimal lead is released into the body. In Greg's case, the high number of pellets increased their surface area and several were lodged in the fluid spaces of his lungs' pleural spaces and his heart's pericardial space, allowing more lead to be dissolved. Also, when shotgun pellets collide with each other

Figure 8.1 Greg LeMond announces his premature retirement from bicycle racing due to mitochondrial myopathy, December 1994.
Photography: Gary Newkirk/Getty Images.

and with bone, they create micro lead fragments, which are hard to see on X-rays but again increase the amount of lead that can be dissolved. When Greg exercises for more than an hour, his blood lead levels rise. Greg was diagnosed with a muscle biopsy which revealed the ragged red fibers that indicate mitochondrial myopathy. Greg's case is very unique because it is rare for world class athlete in his prime to be affected by lead poisoning or have an acquired mitochondrial myopathy. Greg's diagnosis was unusual because most laypeople had never heard of the disease and most doctors at the time had never diagnosed or treated patients with mitochondrial diseases. Greg LeMond's diagnosis made the general public aware of mitochondrial disease and made physicians and researchers more attentive to mitochondrial science.

"We are at a crucial juncture in the perception of these disorders. The recent announcement by the American cyclist Greg LeMond that he is retiring from competitive cycling because of a mitochondrial myopathy has brought these mysterious disorders to the attention of the public.... This announcement has been a seminal event in the public awareness of the mitochondrial disorders."
—Donald R. Johns (1995) [16]

Greg LeMond was not handicapped by his diagnosis and he rose above it, having a successful and active post racing career. He lives like any normal person and still exercises regularly. However, he can no longer compete at the highest levels. He still contributes to the cycling world with innovation in his fitness equipment, watts training devices, a new line of carbon fiber bicycles, sports broadcasting, combating doping and corruption in cycling, and by advancing the science of his beloved sport with this book (Figure 8.1).

References

[1] Gomez C, Bandez MJ, Navarro A. Pesticides and impairment of mitochondrial function in relation with the Parkinsonian syndrome. Frontiers in Bioscience 2007;12:1079–93.

[2] Katan MB, Zock PL, Mensink RP. Trans fatty acids and their effects on lipoproteins in humans. Annual Review of Nutrition 1995;15:473–93.

[3] Kohlmeier L, Simonsen N, van't Veer P, et al. Adipose tissue trans fatty acids and breast cancer in the European Community Multicenter Study on Antioxidants, Myocardial Infarction, and Breast Cancer. Cancer Epidemiology Biomarkers & Prevention 1997;6(9): 705–10.

[4] Zlatohlavek L, Vrablik M, Grauova B, et al. The effect of coenzyme Q10 in statin myopathy. Neuroendocrinology Letters 2012;33(Suppl 2):98–101.

[5] Pintia M, Salomonib P, Cossarizzaa A. Anti-HIV drugs and the mitochondria. Biochimica et Biophysica Acta - Bioenergetics 2006;1757(5–6):700–7.

[6] The American Medical Association. New Policies on Second Day of Voting at Annual Meeting; 2013.

[7] U.S. Department of Health and Human Services. Patient Protection and Affordable Care Act (ObamaCare); 2010.

[8] Grundy SM, Brewer HB, Cleeman JI, et al. Definition of metabolic syndrome: report of the National, Heart, Lung, and Blood Institute/American Heart Association. Circulation 2004;109:433–8.

[9] Mendis S, Puska P, Norrving B, editors. Global Atlas on Cardiovascular Disease Prevention and Control, World Health Organization (in collaboration with the World Heart Federation and World Stroke Organization), Geneva; 2011.

[10] www.MayoClinic.org. Diseases and Conditions: Heart Disease; 2013.

[11] Martin LJ. Biology of mitochondria in neurodegenerative diseases. Progress in Molecular Biology and Translational Science 2012;107:355–415.

[12] www.Cancer.gov. National Cancer Institute at the National Institutes of Health, Fact Sheet: Obesity and Cancer Risk; 2012.

[13] Jones RG, Thompson CB. Tumor suppressors and cell metabolism: a recipe for cancer growth. Genes & Development 2009;23:537–48.

[14] Wang X. The expanding role of mitochondria in apoptosis. Genes & Development 2001;15:2922–33.

[15] www.UMDF.org. United Mitochondrial Disease Foundation, Understanding Mitochondrial Disease: Types of Mitochondrial Disease; 2014.

[16] Johns, Donald R. Mitochondrial DNA and Disease (and correspondence). New England Journal of Medicine 1995;333:638–44.

Slowing the Aging Process

When you are young and vibrant, it is all too easy to take your body and health for granted. But as you get older, you may begin to wonder: how long can you sustain activity and health, what can be expected in terms of physical performance as you age, and what if anything can be done to slow the aging process? Mitochondria are the best way to understand the aging process. The mitochondria theory of aging is the latest and most complete explanation of how we age at the cellular level. This theory is an expansion of the free radical theory of aging. Mitochondria generate but also neutralize many of the harmful free radicals that cause cumulative intracellular damage. Beneficial antioxidant foods work at the mitochondrial level. A healthy diet and lifelong exercise are proving to be the best way to curtail the ravages of aging. Recent studies have shown that exercising your muscles helps your entire body ward off preventable diseases and slows the onset of degenerative diseases in the brain. Although mitochondria are not a mystical fountain of youth, they are the fountain of your energy. Aging cannot be reversed, but understanding mitochondria gives you the opportunity to slow the aging process.

Aging: the human condition

"Our new Constitution is now established, and has an appearance that promises permanency; but in this world nothing can be certain, except death and taxes."
—Benjamin Franklin (1789), in a letter to Jean-Baptiste Leroy

There are few who would refute the truth behind Ben Franklin's witticism, but he probably left out a third certainty, aging. As sure as the rising sun, we get older with each passing day. Aging is part of the universal human condition. It will affect us all and there is no stopping it. One way to cope with the inevitability of aging is to look at it as a grieving process. Grief is how we deal with personal loss and there is no loss more personal than losing yourself while you are still alive. Psychiatry has broken down grieving into a five stage process as a means of guiding people through loss accounting for human emotion but also employing realistic expectation to find a path to final acceptance [1].

Siddhartha

The legend of Buddha was portrayed in Hermann Hesse's novel "Siddhartha." [2] As a young prince, Siddhartha lived in a palace and could have lived out his life in wealth and luxury. However, when he encountered people suffering from aging, illness, and death, he realized that the true human condition is in dealing with these inevitabilities. Through a long journey, he abandoned the trappings of greed, educated himself, realized that suffering was self-imposed, solved the internal grieving process, and reached enlightenment (acceptance).

The Science of Fitness: Power, Performance, and Endurance. http://dx.doi.org/10.1016/B978-0-12-801023-5.00009-0

The five stages of grief (about aging) are:

Denial

When we are young and vibrant, it is all too easy to think that you are invincible and that you are immune from illness and degeneration. Health comes naturally and requires no great effort. At a certain age, the responsibilities of family, career, and finances take priority over fitness, and it is at this stage in life that many young adults stop exercising on a regular basis. However, the best time to prevent aging is when you are still healthy and can begin lifelong behaviors that will maintain your health.

Anger

When athletes get older, training becomes more difficult, results are less and/or come slower, and performance begins to wane. It is natural to become angry or frustrated when your body lets you down with a new back ache, sore knee, slower 10K time, less strength, and less endurance. However, anger and frustration are not constructive emotions. If you understand what happens as you age and know what to expect, you will not be so angry with yourself.

Bargaining

"Maybe if I buy a new lighter bike, it will make up for my riding slower." Although bargaining will not solve the underlying problem, it does acknowledge the problem and tries to seek a compromise. It shows that you are making some progress in trying to find a solution. Sometimes buying new equipment or starting a new sport can motivate athletes and rekindle interest.

Depression

Giving up is a backwards step that needs to be avoided. In the case of exercise, many out of shape, older, or overweight people assume it is too late to start exercising and thus never try. As long as you are cleared to begin exercise by your physician, you can see immediate benefits to your health if you start to exercise at any age, weight, or fitness level.

Acceptance

If you know how the body ages, know what to expect as you get older, and know what you can do slow the aging process, then you can take a proactive approach to getting older. You cannot halt or reverse the aging process. However, with exercise you can function at the level of someone younger than your chronologic age and preserve ample strength and cognitive ability to keep you active and independent for as long as possible (Figure 9.1).

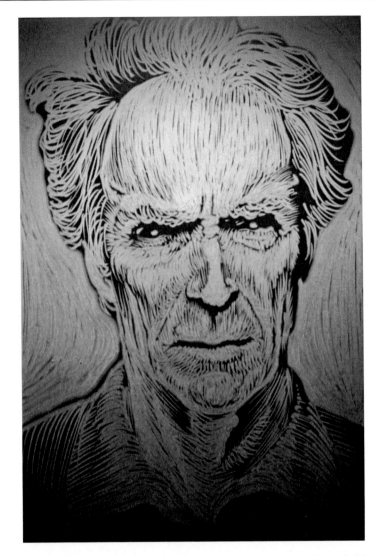

Figure 9.1 During his prolific career, Clint Eastwood was an action movie star in his youth, but won his Academy Awards for producing and directing later in life.
Portrait by Mark Hom.

Physical changes with aging

One of the most noticeable changes in the aging athlete is decreased muscle mass. Age-related muscle loss (sarcopenia) is due to gradual decreases in androgenic hormones (testosterone), decreased neuromotor stimulation, and decreased protein metabolism. Testosterone peaks in young adult men and then diminishes about 1 percent per year after the age of 30 years [3]. There are some conditions such as hypogonadism and "low T" where this drop is more profound and patients may benefit from testosterone therapy [4]. However, we do not advocate supplementary hormones as a "cure" for aging since it can have serious side effects and is banned in competitive sports. As we age, we tend to lose brain cells including the motor neurons that signal our muscles to contract. This decrease in neuromotor stimulation tends to affect the fast twitch muscles the most [5]. Fast twitch muscles are the more powerful and bulkier muscles that develop with weight training. As we will see, muscle mass is a good predictor of longevity, so weight training is a key aspect of fighting the aging process. Amino acids are the building blocks of protein and muscle. There are nine essential amino acids that our cells cannot synthesize and must be included in our diet. There are 12 non-essential amino acids that our cells can synthesize. The problem is that when we age our cells may not be as efficient in amino acid absorption and metabolism and therefore more protein is required in our diet [6]. This problem can be compounded in strict vegetarian diets or caloric-restricted diets. Supplementing your diet with whey protein shakes, when combined with a resistance training regimen, will help maintain muscle mass [7].

The heart can develop problems as we age. The heart muscle can become weaker if we exercise less, the valves may become less efficient, the electrical conduction system can cause arrhythmias, the heart tissue can become scarred with ischemia and infarctions, and the coronary arteries supplying blood and oxygen to the heart can become hardened and narrowed with arteriosclerosis. Heart disease is the leading cause of death, so if you have any chest pain, shortness of breath, exercise intolerance, or are starting a new exercise plan, you should get checked by your doctor. Often cardiovascular disease is insidious with few early symptoms. Assessing your heart attack risk by screening for hypertension, diabetes, obesity, hypercholesterolemia, and behaviors such as smoking is recommended. If your heart or coronary arteries are diseased there are new effective treatments. If you are cleared to begin a fitness program, exercise is one of the best things you can do to prevent heart disease.

As we age, our lungs accumulate damage from pollution, smoke, chemical exposure, and infections. In smoking-induced emphysema, the delicate alveoli are replaced with empty air sacs. In pulmonary fibrosis, the pliable lung tissue becomes stiff and restrictive.

As we age, we lose mineral density in our bones making them weaker, we lose cartilage in our joints causing arthritis and joint pain, spurs grow at our joints and joint capsules resulting in loss of flexibility, and our intervertebral discs lose height making us shorter and stiffer. Arthritis in our joints, injuries and fractures, lowered range of motion, and pinched nerves from spine disease are major causes of disability and can limit exercise participation. In general, low-impact aerobic exercise, stretching, and resistance training are beneficial for bone and joint health.

As we age, we may lose some digestive function and have trouble absorbing some nutrients and vitamins. For example, some older people develop atrophic gastritis that can decrease the absorption of vitamin B_{12}. Having a balanced and varied diet is important as we age and a daily multivitamin may help.

Predictors of longevity

What are reliable predictors of longevity? Although there are accounts of remote villages where several inhabitants live past 100 years, they may be explained by genetics or unusual diets. What we are discussing here are physical measures or behaviors that can predict longevity. Twin studies have shown that longevity is only 20–30 percent genetic [8], which means that 70–80 percent of longevity is due to behavior and lifestyle choices, things we can control. We often perceive an overweight person as being unhealthy, but body mass index (BMI) is actually a poor predictor of longevity. BMI only uses height and weight and does not factor in fitness, strength, or body composition (muscle vs. fat). A better predictor of longevity is muscle mass [9]. This can be more accurately determined with electrical impedance (body fat has more electrical resistance, muscle has less) and abdominal girth. Belly fat correlates with hypertension, metabolic syndrome, lipid/cholesterol abnormalities, type 2 diabetes, and cardiovascular disease [10].

What about personality? We might think of an extroverted, optimistic, unstressed, and laid-back "cool" person as having a longer life expectancy. However, such people tend to have shorter lives because they take part in risky behavior (such as smoking or participating in dangerous activities) and take a less proactive approach to health and diet. It is the quieter more diligent and perseverant well-educated hard worker who has a longer life expectancy, being more likely to hold down a long term job or marriage and being more concerned about health maintenance. A challenging and engaging life is good for your health [11].

One of the most accurate predictors of mortality (impending death) is a simple functional test: how much difficulty a person has lowering themselves to the floor into a seated position and then rising to a standing position unassisted [12]. To perform this maneuver with ease, the person must have core strength, a pain-free back, joint flexibility, balance, and ample muscle mass. These are all qualities that improve with exercise and diminish with inactivity.

Performance changes with aging

What can be expected in terms of athletic performance as we age? Olympic world records are set at a mean age of 26.1 years [13]. Beyond this single average number, there are individuals who peak at a younger or older age and some well beyond this age. However, it is a sobering fact that amongst elite athletes the age range for optimal performance is fairly brief at roughly 26–32 years old. Our cardiac output decreases as we age. Expected peak heart rate in beats per minute is often calculated as 220 minus age, so our hearts do not beat as fast at full exertion as we age. The other component of cardiac output, stroke volume (how much blood is moved per heart contraction) also decreases as we age [14]. In addition, our heart become less responsive to adrenaline (the fight or flight hormone that makes our hearts beat faster and more forcefully) [15].

Figure 9.2 John Muir was a mountaineer and naturalist who helped establish the U.S. National Park System and founded the Sierra Club later in life.
Portrait by Mark Hom.

As noted before, we tend to lose muscle mass as we age. With less cardiac output and less oxygen demand, VO_2 max decreases with age. However, the VO_2 max decline with age in athletes who keep exercising is only half the decline with age observed in sedentary persons [16]. VO_2 max is an estimation of mitochondrial mass and efficiency, so we lose mitochondria activity as we age, resulting in less energy, less speed, and less endurance. But we will minimize mitochondrial loss with age if we keep exercising (Figure 9.2).

One reason for this performance drop off can be something we can control, training time and intensity. As we get older, we have more responsibilities and our priorities change, often resulting in less training time. With arthritis and/or accumulated injuries, we may have pain or need more time to recovery between training sessions. The net result is that we tend to have less time to train and find excuses not to exercise. One advantage of mitochondrial based training is that it favors intensity (quality) over distance and time (quantity). In a shorter period of time, it is possible to get a more stimulating workout that fits into a busy schedule and can avoid chronic overuse injuries. Although some diminution of performance is to be expected, it does not mean that we should give up. We need to stay motivated and keep exercising, albeit with more realistic goals and expectations.

The free radical theory of aging

Aging is a slow, irreversible, and cumulative process. Harman noted that organisms with a high metabolic rate (increased oxygen consumption) had a shorter life span and that hyperbaric oxygen and radiation caused free radical oxygen damage to cells [17]. As noted previously, oxygen is highly reactive and wants to combine with other chemicals. There are other forms of oxygen with a single unpaired electron that are super reactive and more damaging such as superoxide (O_2^-), hydroxyl radical ($\cdot HO$), and hydrogen peroxide (H_2O_2). When these reactive oxygen species are created by the metabolism inside cells, these free radical chemicals can damage vital cell components on contact, damage DNA, and impair cell processes. This free radical theory explained how aging could be a gradual yet cumulative process inside our cells, related to oxygen metabolism. Slowing metabolism with caloric restriction to decrease free radical damage showed some promise [18], but is a difficult lifestyle to follow and can lead to reduced muscle mass. The hope was that aging could be slowed or halted with antioxidant nutrient or vitamins. However, experiments failed to show significant extension of life with antioxidant therapy.

The mitochondrial theory of aging

More evidence pointed to mitochondria as being central to the aging process. As cells age, they lose mitochondrial energy with decreased ATP production. Mitochondrial DNA (mtDNA) is less protected (lacking the histones seen with nuclear DNA) and more vulnerable to free radical damage and mutation. Mitochondria are at the center of cell metabolism and are the main generators of free radicals. Mitochondria are suspected of having an important role in neurodegenerative diseases such as Alzheimer's disease and Parkinson's syndrome [19]. As mitochondria decay with age, they leak more of the free radicals they generate. Aging can then be seen as a vicious cycle of aging mitochondria accumulating mtDNA mutations and releasing more free radicals. With these considerations, the free radical theory of aging evolved into the mitochondrial theory of aging [20]. This is currently the most accurate model to explain the aging process. However, it is not a simple theory. Free radicals and mitochondria strike a balance. During exercise, free radical production is one of the triggers that promotes mitochondrial biogenesis (increasing numbers of energy-producing mitochondria) [21]. Mitochondria also have efficient means of handling free radicals with internal antioxidant systems (such as glutathione) [22]. When exercising to the point of exhaustion, these antioxidant systems can be overwhelmed and tissue damage can occur [23]. However, it is a mistake to think that exercise accelerates aging. Exercise done properly and in moderation is one of the best things we can do to slow aging.

Figure 9.3 Theodore Roosevelt (26th U.S. President) led the Rough Riders, was a rugged outdoorsman, and stayed active his entire life.
Portrait by Mark Hom.

How to slow the aging process

Although aging is an inevitable process and an unavoidable part of the human condition, understanding how it works and what to expect makes aging easier to deal with. There are things we can do to slow aging, making us feel younger than we are. We can make proper lifestyle choices when we are still young by following a lifelong exercise plan and healthy diet. With exercise we can build or maintain muscle mass, improve coordination and balance, and preserve mobility and independence. Exercise is helpful in preventing numerous diseases including hypertension, metabolic syndrome, obesity, type 2 diabetes, colon and breast cancer, fractures and injuries, and brain degeneration [24]. With a proper diet, we can support protein synthesis and control caloric intake. By being proactive, we can make our mitochondria work for us in generating high energy levels and maintaining fitness as we get older (Figure 9.3).

References

[1] Kubler-Ross E. On Death and Dying. London: Routledge; 1969.

[2] Hesse H. Siddhartha. New York: New Directions; 1951.

[3] Feldman HA, Longcope C, Derby CA, et al. Age trends in the level of serum testosterone and other hormones in middle-aged men: longitudinal results from the Massachusetts male aging study. The Journal of Clinical Endocrinology and Metabolism 2002;87(2):589–98.

[4] Bain J. Testosterone and the aging male: to treat or not to treat? Maturitas 2010;66(1): 16–22.

[5] Lexell J. Human aging, muscle mass, and fiber type composition. Journals of Gerontology Series A: Biological Sciences and Medical Sciences 1995;50:11–6.

[6] Fujita S, Volpi E. Amino acids and muscle loss with aging. Journal of Nutrition 2006; 136(1 Suppl):277S–80S.

[7] Phillips SM, Tang JE, Moore DR. The role of milk- and soy-based protein in support of muscle protein synthesis and muscle protein accretion in young and elderly persons. Journal of the American College of Nutrition 2009;28(4):343–54.

[8] Hjelmborg J, Iachine I, Skytthe A, et al. Genetic influence on human lifespan and longevity. Human Genetics 2006;119(3):312–21.

[9] Srikanthan P, Karlamangla AS. Muscle mass index as a predictor of longevity in older adults. American Journal of Medicine 2014;127(6):547–53.

[10] Klein S, Allison DB, Heymsfield SB, et al. Waist circumference and cardiometabolic risk: a consensus statement from Shaping America's Health: Association for Weight Management and Obesity Prevention. American Journal of Clinical Nutrition 2007;85(5):1197–202.

[11] Friedman HS, Martin LR. The Longevity Project: Surprising Discoveries for Health and Long Life from the Landmark Eight-Decade Study. New York: Plume; 2012.

[12] Brito LBB, Ricardo DR, Araújo DSMS, et al. Ability to sit and rise from the floor as a predictor of all-cause mortality. European Journal of Preventive Cardiology 2014;21(7):892–8.

[13] Berthelot G, Len S, Hellard P, et al. Exponential growth combined with exponential decline explains lifetime performance evolution in individual and human species. Age 2012;34(4):1001–9.

[14] Ogawa T, Spina RJ, Martin WH 3rd, et al. Effects of aging, sex, and physical training on cardiovascular responses to exercise. Circulation 1992;86(2):494–503.

[15] Seals DR, Taylor JA, Ng AV, Esler MD. Exercise and aging: autonomic control of the circulation. Medicine and Science in Sports and Exercise 1994;26(5):568–76.

[16] Rogers MA, Hagberg JM, Martin WH 3rd, et al. Decline in VO2max with aging in master athletes and sedentary men. Journal of Applied Physiology 1990;68(5):2195–9.

[17] Harman D. Origin and evolution of the free radical theory of aging: a brief personal history. Biogerontology 2009;10(6):773–81.

[18] Ungvari Z, Parrado-Fernandez C, Csiszar A, de Cabo R. Mechanisms underlying caloric restriction and lifespan regulation: implications for vascular aging. Circulation Research 2008;102(5):519–28.

[19] Federico A, Cardaioli E, Da Pozzo P, et al. Mitochondria, oxidative stress and neurodegeneration. Journal of the Neurological Sciences 2012;322(1–2):254–62.

[20] Loeb LA, Wallace DC, Martin GM. The mitochondrial theory of aging and its relationship to reactive oxygen species damage and somatic mtDNA mutations. Proceedings of the National Academy of Sciences 2005;102(52):18769–70.

[21] Yoboue ED, Devin A. Reactive oxygen species-mediated control of mitochondrial biogenesis. International Journal of Cell Biology Epub 2012;Epub 2012 May 30.

[22] Marí M, Morales A, Colell A, et al. Mitochondrial glutathione, a key survival antioxidant. Antioxidants & Redox Signaling 2009;11(11):2685–700.

[23] Sachdeva S, Davies KJ. Production, detection, and adaptive responses to free radicals in exercise. Free Radical Biology and Medicine 2008;44(2):215–23.

[24] Assistant Secretary for Planning and Evaluation. Physical Activity Fundamental to Preventing Disease. Washington, DC: U.S. Department of Health and Human Services; 2002.

Gauging Fitness

10

Measuring physical performance

Athletes at all levels like to see results, whether it is a steady improvement in fitness, a personal best time, more endurance, more speed with less effort, a slimmer physique, or beating a rival to the line. In this section, we will describe the different methods an athlete can use to gauge fitness, the advantages and disadvantages of different measurements, the trappings of relying only on one type of measurement, measurements which encourage improvement, electric gadgetry that can be useful to the athlete, the importance of measuring power, and how to be more in tune with your body to gauge fitness without any measurements at all. In ancient times, before the advent of chronometers, speedometers, bicycle computers, and wattage sensors, athletes relied on competition to gauge fitness. The ancient Greek Olympics were based on the principle of fair play and sportsmanship to determine who was the fastest, the strongest, and the best. The ultimate gauge of fitness at that time was competition. Athletic games created champions on par with war heroes with tales of athletic feats recorded by Homer in the Odyssey [1]. For many, competition is still the most challenging and fascinating aspect of athletic endeavor. If a car race was determined solely on engine horsepower alone, the results of a NASCAR race could be determined in the laboratory/garage on a dynamometer. But a NASCAR race is won with skill, tactics, risk taking, and smooth application of power. Spinning the wheels at full throttle would be a recipe for disaster.

But what about fitness athletes? Why do they spend hours training when winning a competitive race is of little importance? Even they want to see improvement, and in this case they are competing with themselves, to outdo their former selves. In lieu of competition, gauging fitness is even more important when training for health and vitality. Athletes want to know their level of fitness and how it can be improved, or as we age how fitness can be maintained at high levels.

The different types of fitness gauging will be divided into the following categories:

The Science of Fitness: Power, Performance, and Endurance. http://dx.doi.org/10.1016/B978-0-12-801023-5.00010-7

Newtonian physics

Although athletic activity (at its best) is a fluid and dynamic experience that seems to have little to do with equations and mathematics, when we put our bodies in motion, we become part of a physics experiment. Sir Isaac Newton applied basic measurements of distance, time, speed, and acceleration to describe the movement of objects [2]. He also invented calculus to explain the mathematics of change and to explain inertia, force, and gravitation. Although this will be a calculus-free explanation of exercise, Newtonian principles and units of measurement are useful to athletes. Physics is the study of objects in motion, so it is important to understand some of the principles of physics when we put our own bodies in motion during exercise. To best understand these principles, it is also important to have an intuitive sense of what they mean, just as Isaac Newton and Albert Einstein used simplification and common observations to make their theories intuitively logical.

"Everything must be made as simple as possible. But not simpler."

—Albert Einstein

Distance

Distance is an important measure for athletes. Olympic track and field events are defined by precise distances which select athletes who are most adept at short sprints, middle distance running, or endurance events. Distance defines what athletes can expect on a training ride or run. As a beginner in a new sport, training distances start short and generally extend once fitness improves. As we train our energy efficiency improves with boosting of our mitochondrial mass and improved energy metabolism, allowing us to extend the training session. So it is natural to increase training distance during a season of training. Often long distance becomes a year's training goal, such as a 26.2 mile marathon race or a 100 mile "century" bike ride. By logging distance, an athlete can see improvement in the effectiveness of training and in fitness. However, there are some downsides to relying on distance too much. Mileage says nothing about the pace or intensity of a workout. A 5 mile leisurely walk around the lake will not build as much muscle nor burn as many calories as a fast paced 5 mile run. If distance is the only goal, some athletes can fall into the trappings of overtraining. There is some evidence that ultra-long distance training can have negative effects on health because of free radical damage [3]. There is also the issue of overuse injuries that can end up decreasing training time. Ultra-long distance training favors slow twitch muscles and endurance at the expense of fast twitch muscles and power. Muscles need recovery time and dietary protein to rebuild from the stresses of exercise. Unless nutrition is sufficient and properly timed and unless adequate rest is allowed, muscle loss can occur with overtraining. If preparation for a long distance event is the primary training goal, there can be a sharp drop off in training after the event. Sometimes runners are heard saying, "I haven't done much running since the marathon." Part of this phenomenon is the real need for recovery time but there is also the need to keep motivated after a goal is accomplished. It is never a good idea to stop exercising for long periods. We must have a plan for the time after we have achieved our goal or, better yet, have a long term plan with many goals and new physical challenges.

Time

Time is helpful to athletes in several ways and is easy to measure. The typical starting line photo has long distance runners activating their stop watches. If athletes train or race at set distances, time tells them a lot about their performance and fitness level. Early in the season, it is easy to record and graph shorter and shorter times over a set distance and see real improvement (Figure 10.1). However, just relying on course time can have its negatives. Runners are notorious for running the same fixed course every training session, in order to get the most accurate time recording. What they may find is early steady improvement (as the body, cells, and mitochondria adapt), but then a plateau in fitness as the same distance and topography offers no new challenge and no physical adaptation is required.

Obsession with distance and time

"At one time I was intrigued with target archery for the benefits of strengthening less-used back muscles, the mental focus, and the technical equipment aspects of the sport. I set up my target in what I though was an unused corner of an empty athletic field with a large steep slope as my safety backstop. At full draw, I noticed two runners come into my peripheral vision. Apparently that far corner of the field was part of their training run. Without missing a step, they ran between me and my target, rather than shorten their run by a few feet. The experience taught me how obsessed runners can be with their training distance."

—Mark Hom

Figure 10.1 Road race along the James River in Richmond, Virginia.
Photography: Alan Cooper.

Time can be used in another useful way. In today's hectic world, when training must be scheduled to balance with work time and family time, we are often left with a set amount of time to train. Time is also the main stipulation in US government recommendations for exercise: 2.5 to 5 hours of moderate exercise per week (for adults). The recommendations also say that with vigorous exercise only half the time is needed. Therefore, when strapped for time, it is beneficial to train vigorously. This sort of schedule will improve strength, endurance, and mitochondrial metabolism. Another solution is to replace idle time (car commuting) with exercise time by bike riding or walking to work. The more miles we drive each day, the less healthy we become. Driving can be aggravating and irritating causing the release of stress hormones (such as cortisol) that tax the body and mind and lead to weight gain. Fight or flight hormones (such as adrenaline) are released into your blood stream when you are cut off or honked at, but cannot be burned off when you sit trapped in a car seat [4]. Bicycle commuting allows you to convert this frustrating idle time into exercise time, improving health and allowing more family time (when you get home your workout is done). Going all out for an hour is quite a challenge. Over a long and storied career, Eddy Merckx claimed that the most physically taxing events were his one hour record attempts (Figure 10.2).

Figure 10.2 Joop Zoetemelk (Gitane) and Eddy Merckx (Molteni).
Photography: The Nationaal Archief in The Hague.

Rubber band legs

When I was in high school, after school team sports were not an option. Both of my parents worked full time and could not chauffer me to play sports. It was up to me to fit exercise into a non-car dependent schedule. I would generally finish my homework in the two hours immediately after getting home from school, leaving me with a couple hours of daylight to ride my bike. Not employing a formal training system, I would simply ride as hard as I could until sunset. As I got stronger and faster I would seek out the steepest hills in my area. When I would get home I was completely spent and my legs would feel like rubber bands when I carried my bike down to the basement. I later started bike commuting to school, which was a long boring bus ride and a distance most students thought was too far to be practical. Assisted by adolescent hormone surges, this was when I experienced the greatest improvement in my fitness.

—Mark Hom

How much distance and how much time is optimal for exercise? This is highly dependent on the individual, but there are minimums for the body to do enough work and burn enough calories, just as there can be overtraining where the benefits begin to taper off. The mind is not very good at estimating distance or speed. However, we are better at estimating time since our brains run on circadian rhythms. Therefore, we often know intuitively how much time we need to exercise and when we have had enough. If you make the best use of time, you cannot go wrong.

Speed

Speed is distance divided by time. Speed defines the pace of group rides or runs and is an important measure of performance improvement. If you can travel the same course in less time you are training faster and improving. If you can travel greater distance at the same average speed, you are also improving (endurance). There are some limitations in relying just on speed. If you head out on a bike ride into a stiff head wind, your speed will be less compared with a calm day, yet you will exert more effort from the increased wind resistance and benefit more despite riding slower. Similarly you can have a tremendous workout climbing a steep hill or mountain at a relatively slower speed compared with a flat course. The force of gravity will also allow you to speed downhill; the rush of air and sensation of velocity are nice rewards for enduring the climb. When the pace of a group ride is dictated by speed, there may be some members who want to maintain the same speed throughout the ride and get upset when the pace quickens. If you feel capable of going faster you should not feel restricted by an arbitrarily set pace. Greg LeMond would often eschew his team training rides and ride solo instead. He could push himself when he was feeling strong and not waste his time maintaining a slower pace. Bicycling computers and GPS make it easy to follow speed and maintain a steady pace. Just be aware that to break through to higher levels of fitness, you have to break the pace sometimes.

"It never gets easier..."
One of Greg LeMond's most famous quotes about training is "It never gets easier, you just go faster." Greg was responding to the misconception that at some high level of elite competition, training becomes easy. Even at the highest of fitness levels, it never gets easy. If you are always pushing yourself to improve, training is always a challenge. But what you can expect and what you will be rewarded with is more speed.

Acceleration

Many races are won at the end with a finish line sprint. Not only must the athlete endure to the end, there has to be enough power to "kick" to the line. The key to winning a sprint is to accelerate. Acceleration is speed over time. Acceleration requires much more energy because we are fighting inertia. Inertia is what keeps objects at rest or at a constant speed. One of the best ways to conserve gasoline when driving a car is to accelerate slowly from stops and avoid "jack rabbit" starts. That is because hard acceleration takes more power and energy. To make good use of this fact, use acceleration in training: accelerate from stop signs to cross intersections faster, and sprint to break up a steady pace. Training against this increased resistance will make you stronger. Bicycle computers generally do not record acceleration even though it is a simple calculation (change in speed over time). Perhaps we do not need to be told as we can certainly feel it when it happens. One of the most exciting moments in a race is the breakaway, when a solo rider accelerates away from the steady pace of the main group. Acceleration is also the key to bridging gaps and chasing others down. Being strong enough to accelerate and knowing when to do so will help you win races.

Mass

Mass (or weight) is a simple way to change resistance when training. A few definitions first: the mass of an object does not change, but weight can vary depending on gravitational forces. For situations on the surface of Earth we will use the two terms interchangeably. A body builder will reach for heavier weights when training to increase muscle mass. The greater gravitation pull on a heavier weight causes more muscle fibers to engage during the lift, triggering muscles to hypertrophy. To remain buoyant, swimmers do not add weight, but they can add fins, paddles, or gloves to increase resistance during the power stroke when training. In competitive swimming, a slight finger spread adds propulsive force to the stroke [5].

"You don't need a new bike"

My first multi-speed bike was a heavy steel-framed touring bike that weighed nearly 40 lbs. When I decided to spend my teenage life savings on a new light-weight bike my father said, "You don't need a new bike. You'll get a better workout with your old heavy one." In a way he was right. Climbing hills on that heavy bike was a real challenge but made me stronger. So why do cyclists pay thousands of dollars to have a bike that is a few pounds lighter? A light-weight bike is more responsive and pleasurable to ride. It will accelerate and climb faster with the same effort. There is a more rewarding sensation of speed. A light-weight carbon fiber bike can be tuned to have just the right amount of give and stiffness to be efficient yet also comfortable. And a lighter bike will be more competitive. Sometimes it *is* about the bike.

—Mark Hom

Altitude

Climbing is another way to increase resistance when training. In this case, we are using our own body weight and gear as resistance against gravity. Although novice athletes may dread the agony of climbing, ascending hills and mountains is a sure way to get fit against the unrelenting pull of gravity. Einstein made the connection between gravity and acceleration in his general theory of relativity, theorizing that a person in a windowless space ship accelerating at a rate of 1 g (9.8 m/s^2) would not be able to tell that he is not on Earth. Gravity and acceleration are the same [6].

Climbing stores energy. When we are at the bottom of a valley, our potential energy is low. As we climb to a summit our potential energy increases and is stored in our elevated position. As we descend, that potential energy is converted into motion

Figure 10.3 Photo of Greg descending in full tuck, converting potential energy into kinetic energy. Photography: Pascal Pavani, AFP/Getty Images.

(kinetic energy) and we speed down the mountain. Riding in the Sierra Nevada Mountains in his youth, Greg LeMond became adept at racing up mountains. As a professional racer, he also showed great skill in rapid descents, often making time on his opponents by taking risks during high-speed alpine mountain descents. The secret was to using gravity to his advantage by maintaining momentum in the turns, using an aerodynamic tucked position, and braking as little as possible (Figure 10.3).

Total climbing altitude during a ride can be calculated by some cycling computers by measuring changes in barometric pressure. Another simpler way is to train using hill repeats (explained in Chapter 11: BEAST Fitness Training). The more hill repeats you complete, the more climbing you have accomplished.

Force

Force is Mass times Acceleration. We can apply more force by moving more mass or accelerating harder. Applying force when training makes us stronger as our bodies adapt to the new challenge. Since acceleration is equivalent to gravity, gravitational force is another form of resistance we can train against. Mechanical work is force times distance. It takes work to climb a mountain and mechanical work is the gravitation force of our body (and equipment) times the height climbed. Torque is a type of work, which is force applied over a lever arm (or twisting force). When we pedal a bike, our legs apply torque to the crank arms, driving the bike forward. Sensing this torque (with electronic strain gauges) is the principle behind bicycle power meters. Power is Work over Time. James Watt experimented with building ever more efficient steam engines and he needed a way to calculate how well steam engines functioned. He defined a horsepower by how quickly a horse could lift a prescribed weight a certain height [7].

Work versus power

Physics students sometimes confuse work with power, so it is important to have an intuitive sense of what these two terms mean. Work is force times distance. Power is work over time. Let's say a manager of a store tells his night-shift employees to restock the shelves. This involves lifting objects against the force of gravity onto shelves of various heights (force times distance). The next morning the manager sees the restocked shelves as evidence that work was done. A week later, the manager buys a fork lift and the night-shift reports that this time the restocking (the same amount of merchandise) took only half of the time. The same amount of work was accomplished but faster. This is because more power was used.

When the Peloton reaches the summit of the Alpe d'Huez climb in the Tour de France, each rider has accomplished work, but it is the rider with the best power to weight ratio who wins the stage.

As we will see in a following section, power is the best measure of athletic performance. The more power an athlete has, the faster the athlete can climb, sprint, and accelerate. Wattage training can improve fitness. Power is what wins races (Figure 10.4).

Physiology

Unlike an inert metal sphere in a Newtonian physics experiment, the human body is a dynamic, innervated, and responsive living organism that readily adapts to motion and exercise. There are several physiologic measurements that change during exercise and are of value to athletes.

Heart rate

The most dramatically changing measurement during exercise is heart rate. The heart may beat as low as 50 beats per minute in a well-trained endurance athlete at rest, to over 200 beats per minute during highly intense exercise (four times faster). The heart rate also changes depending on the health of the individual, during times of shock, major bleeding, severe anemia, dehydration, heart arrhythmias, and mental stress. When a doctor feels your radial pulse, the doctor takes a quick assessment of your overall condition by the pulse's strength, rate, rhythm, and regularity.

The normal resting heart rate should be between 60 and 80 beats per minute (bpm). Sometimes up to 100 bpm is acceptable in the doctor's office because of "white coat syndrome." A patient's heart rate and blood pressure will often rise at the sight of a physician in a white lab coat. The best way to measure your resting heart rate is when you first wake up, are calmly lying in bed, and are not nervous nor have the strong urge to urinate. The rate can only go so low because the body at rest still needs blood flow to keep vital organs such as the brain, heart, liver, and kidneys adequately perfused. With training some athletes can drop their resting heart rate below normal to the 50s and even 40s. That is because their hearts are pumping so efficiently, that they move

Figure 10.4 The LeMond Revolution with Watt Box gives a true road feel during indoor training and records the cyclist's power output.
Photography: Corey Gaffer/GregLeMond.com.

enough blood at low heart rates. The maximum heart rate in healthy adults is often calculated (somewhat arbitrarily) as 220 minus age, so for a 30 year old person, a maximum heart rate of 190 bpm can be expected [8]. Between the extremes of say 70 and 190 bpm is where changes in heart rate can indicate how vigorously an athlete is exercising. Entire books have been written about heart rate training, dividing heart rate ranges into various (somewhat arbitrary) zones, and designing training regimens centered on these zones. The principle is to better inform the athlete when they are training hard enough to get benefit, when they are burning the most calories, and when they are going slow enough to recover. Heart rate can tell you more than group pace because different individuals in the group may have to work harder than others to maintain the group pace. Heart rate has become easier to measure. Before, uncomfortable chest strap sensors and dedicated recording devices were required.

Now the technology has been miniaturized to self-contained wrist bands "watches" or fingertip devices that plug into a smart phone.

The heart rate changes to adapt to the demands of muscles during exercise. A faster heart rate will deliver more oxygen to our muscles (remember that it is our mitochondria that consume our oxygen), deliver more fuel (namely glucose), and carry away CO_2 and lactic acid. It was once thought that the heart rate changed mainly in response to fight-or-flight hormones such as adrenaline (aka epinephrine). When adrenaline is released into the blood stream it increases blood flow and oxygen to muscles by increasing heart rate, heart contractions, and airway dilatation, and by shunting blood away from organs such as the intestines. The adrenaline effect is what allows a mother to lift an extremely heavy object off of her child in an emergency situation, and is a part of our survival evolution. However, as most athletes know, fear and panic are not prerequisite to outstanding physical performance. Well-trained athletes ramp up their muscle power, but do so in a smoothly controlled fashion.

Our hearts are controlled by a subconscious autonomic nervous system consisting of opposing yet complementary (Yin Yang) balance: the sympathetic and parasympathetic nervous systems. The sympathetic nervous system is more primitive and resides along our spinal column and can respond to urgent action and reaction (fight-or-flight) with a faster heart rate and stronger heart contractions. The parasympathetic system resides in our brain stem (still below conscious control) and has a calming effect and controls nonurgent bodily functions. The autonomic nervous system gets input from our senses, emotions, blood pressure, CO_2 levels, O_2 levels, and blood acidity and then conveys this information to the intrinsic conduction system in your heart. Before severely diseased cardiac patients receive a heart transplant, their diseased hearts cannot respond adequately to the autonomic nervous system and they generally require a cardiac pacemaker to make their hearts more functional. After a heart transplant, all nerves going to and from the heart are cut by the surgeon and the pacemaker is removed or deactivated. What happens when these heart transplant patients exercise? A transplanted heart is not paralyzed (as would be a leg muscle) because the intrinsic conduction system will keep the heart beating spontaneously. The post-transplant heart rate will increase with exercise and decrease when at rest, but the response is significantly delayed compared with a normal native innervated heart. Instead of direct nervous control the transplanted heart responds to adrenaline release into the blood stream (the adrenal glands are controlled by the autonomic nervous system) [9]. With time, some of the autonomic nerves can regrow and restore some degree of autonomic control to the transplanted heart.

As we will see in a following section, well-trained athletes become in tune with their bodies and can sense with some degree of accuracy their level of exercise intensity. Although we may not intuitively sense our exact heart rate, more importantly we can sense how hard we are working using several clues, to be discussed later in self-perception of effort.

Cardiac output

Perhaps even more important than heart rate, but much harder to measure is cardiac output in response to exercise. The heart is a hard working muscle, tasked with delivering oxygen and fuel by pumping action. How much blood the heart pumps is cardiac output, which is heart rate $\times$ stroke volume. As the heart becomes conditioned to increasing levels of exercise, it becomes a more efficient pump with more blood moved with each stroke. Not only is the cardiac muscle getting stronger, the mitochondria generating the energy become more numerous. Under the microscope, mitochondria can take up to 1/3 of the volume of heart tissue. A constant supply of energy is needed because the heart cannot take a day (or even a brief moment) off the job.

Another way the stroke volume increases is with the Frank-Starling effect [10]. When exercising, more blood returns to the heart and distends the heart chambers. This stretch filling of the heart causes the heart to contract more forcefully and pump more blood with each stroke. Although stroke volume and cardiac output are very important, they require sophisticated medical imaging, and even then only when the body is at relative rest. As we will see later, VO_2 max is an indirect measure of cardiac output and a very good measure of athletic power.

Respiratory rate

Similarly linked with the autonomic nervous system and connected to the cardiovascular system, is the respiratory system. As we exercise our respiratory rate increases to capture more oxygen (O_2) from the air and to eliminate more carbon dioxide (CO_2). Mitochondria consume the O_2 and create CO_2 as they convert fuel into useable cell energy. Sensing CO_2, O_2, blood acidity, temperature, and blood pressure, the autonomic nervous system tells us when we need to breathe faster. Unlike heart rate, we have some conscious control. Breathing is autonomic when we sleep, but when we are awake we can control our breathing and hold our breath when needed. However, there is a strong urge to breathe more when we train hard. In the case of cold water submersion, the urge to breathe is suppressed and metabolism is slowed in the "diving reflex."

Belly breathing

Not only can we consciously control how rapidly we breathe, we can control how deeply we breathe as well. Many athletes do not breathe properly and thus do not take full advantage of their lung capacity. Watching video of Jan Ullrich (nicknamed the "Big German") during epic Tour de France climbs, his technique was to remain seated and breathe deeply. From the side, one could observe his abdomen bulging during each inhalation. He was using his large diaphragm muscles to maximize his lung capacity, whereas some athletes rely too much on small rib cage muscles. Belly breathing may not look attractive, but is the most efficient way to breathe. Such breathing technique is a fundamental basis of the Zen arts and body control.

As important as breathing is to athletes, it is not included in consumer sports monitoring devices. That is because the athlete can readily sense and control their own respiratory rate. That leaves the responsibility to the athlete. Not only must the athletes breathe faster and more deeply during high intensity activity, breathing can also preload oxygen into muscle myoglobin before a sprint or climb and breathing must be maintained after high exertion to hasten recovery.

Blood pressure

High blood pressure (i.e., hypertension) is often called the "silent killer" because it may have no symptoms, but in the long run can increase risk of heart attack (myocardial infarction), stroke (cerebrovascular infarction), and end organ damage (such as renal failure). Measuring blood pressure with an arm cuff is not very practical when training on the road, but it is important for athletes to know their resting blood pressure because hypertension is treatable. Because of "white coat" syndrome, your blood pressure at the doctor's office may be abnormally high due to anxiety [11]. A digital home blood pressure cuff is not very expensive and allows you to try biofeedback (self-calming) to see if you can reduce the anxiety effect and have a normal blood pressure at rest (120/80 mm Hg).

The good news for athletes is that regular exercise can lessen hypertension (often as much as medication) and in some cases allow hypertensive patients to come off of their blood pressure medication. The beneficial exercise effect is threefold: the trained heart becomes more efficient and does not have to work as hard to move the same amount of blood, the arteries relax and become more supple to lower peripheral resistance, and arteriosclerosis (hardening and narrowing of the arteries) is prevented [12].

Blood pressure is moderated by the autonomic nervous system and can change with dehydration, anxiety, emotion, shock, exposure, salt imbalance, and hormonal abnormalities. Dizziness, nausea, lightheadedness, and fainting can be signs that an athlete's blood pressure is too low which can happen with inadequate hydration when training too hard in hot weather.

A person's resting blood pressure tends to go up when around other people and tends to lower when petting a dog [13]. The theory is that humans are judgmental and trigger anxiety, whereas dogs are nonjudgmental and allow people to relax. A dog's blood pressure will lower when petted and rise when the petting stops. This shows that across different species that blood pressure is influenced by complex emotions.

Cadence

In cycling cadence is how frequently your legs turn the cranks in rotations per minute (rpm) and in running cadence is how frequently your feet hit the ground in strides per minute (spm). Since a cycling rpm counts both feet in one complete turn of the crank, a cycling cadence of 80 rpm is the same as a running cadence of 160 spm. In both cases, the cadence can be affected by crank arm length or stride length, respectively. Swimmers move their arms at a lower cadence in the range of 45–65 spm, mainly due to the increased resistance of water. Cadence is under voluntary control and is a major way athletes change pace. Because muscles have a contraction phase and a relaxation phase, there tends to be a sweet spot or range of optimal cadence. Cadence can be momentarily lowered during times of high power output such as short hill climbing, and momentarily raised during times of acceleration and sprinting. In bicycle time trialing, where sustained high-powered output is required, there is a modern trend for higher cadence. Because power is torque on the pedals multiplied by the rpm, high levels of power can be generated with a high cadence, yet the muscles do not pump up and restrict arterial flow. Muscles are contained inside sheaths and if cadence is too low and foot pressure too high, the muscles can swell and restrict arterial blood flow into the muscle sheath compartment. In severe cases, muscle damage can occur (compartment syndrome). The optimal cadence range is individual and depends on body geometry, crank/stride length, and fitness. A new cyclist cannot just jump on a bike and pedal efficiently at high cadence because this technique requires a strong heart and efficient cardiovascular system that comes with long term training. Riding, running, or swimming at higher cadence can be a both a training goal and performance measure. A magnet affixed to a crank arm and a frame-mounted sensor allow cadence to be recorded on a bike.

Pulse oximetry

With all of our discussions about oxygen delivery and oxygen consumption, wouldn't it be nice to have the technology to measure oxygen saturation in the blood? In fact, this technology already exists and is used in every modern hospital to monitor critically ill patients and people undergoing surgery and medical procedures. Oxygen desaturation can be one of the first indications that a patient is in trouble, so monitoring oxygenation is very helpful in critical hospital situations. When red bloods cells are fully saturated with O_2 they are bright red. This redness can be measured with a fingertip clamp sensor that monitors and records both pulse (heart rate) and oxygen saturation (O_2 sat). This technology has recently been miniaturized and is currently available to connect with iPhones and Android smart phones (Figure 10.5).

We tested the Masimo iSpO$_2$ pulse oximeter and found it to be very easy to use and very accurate. The simultaneously recorded heart rate data and O_2 saturation data could be stored and graphed for later analysis. These are some of our preliminary findings:

A person with a healthy heart and lungs could match the muscle O_2 demands during exercise with a higher respiratory rate (normally CO_2 driven) and higher cardiac output. That is how marathon runners can run for hours and not become cyanotic. To cause a measurable drop in O_2 sat, a healthy test subject had to run up stairs while performing short breath holds to drop down from 99 percent to 95 percent, which is not a marked change. The graphing function allowed us to see this small change, as just glancing at the O_2 sat number would not have been that telling. Similar changes were detected during fast sprinting with lower saturations noted when there was no warm up, steady high saturation during low level and moderate exercise, a saturating effect with hyperventilation, and return to high saturation during active recovery from high intensity. There may not be broad use of pulse oximetry for healthy athletes as they could stay between 99 percent and 95 percent all day. On the other hand the combined heart rate and O_2 sat data was fascinating from a physiologic perspective when you understand what was causing the subtle changes, such as hyperventilation and preloading myoglobin with O_2, the effect of adequate warm up, what it takes to significantly desaturate (super high intensity in healthy athletes), and how quickly saturation recovers to normal levels. The fingertip clamp is less bulky than the chest strap many athletes wear just to get heart rate data. Perfusion index correlates with capillary refill, but is measured at the fingertip and therefore does not directly measure muscle perfusion.

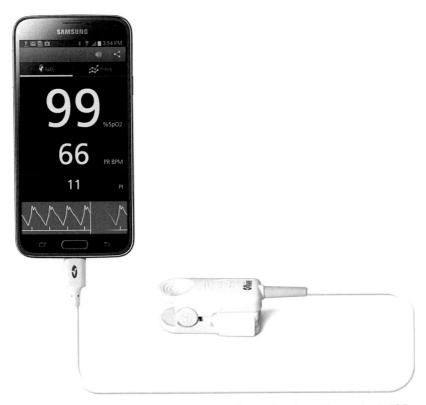

Figure 10.5 The Masimo iSpO$_2$ clamps onto your finger, plugs into either an Apple IOS device or Android device (different versions), and records pulse oximetry, pulse rate, and perfusion index.
Photography: Masimo Corporation.

Here are some possible niche uses of the Masimo iSpO$_2$ for specific types of athletes and exercise situations:

1. High altitude training: The US Olympic training camp is in Colorado Springs well over a mile above sea level. Living at high altitude causes some adaptive changes to low pO$_2$ such as increasing hemoglobin. Pulse oximetry could help monitor for O$_2$ desaturation and detect those at risk for altitude sickness when exercising intensely at the lower pO$_2$ found at high altitude.
2. Fitness trainers who train their athletes at altitude or push them very, very hard: Trainers like to have data and might be interested in even small changes in O$_2$ saturation. The iSpO$_2$ makes this data very precise and repeatable.
3. Extreme mountain climbing: where high altitude, high exertion, hypothermia, and sometimes supplemental O$_2$ combine to make O$_2$ a key element of performance and safety.
4. Patients with emphysema and chronic obstructive pulmonary disease (COPD): It is well known that some COPD patients are CO$_2$ retainers and have their respiratory rate controlled more by O$_2$ levels in the blood than CO$_2$. We also know that they can desaturate faster and lower than normal people. Too much supplementary O$_2$ can suppress respiration and can be

a bad thing. Just because someone has COPD does not mean that they cannot exercise. In fact, exercise can improve many respiratory conditions. The $iSpO_2$ might help people with COPD/emphysema exercise within safe parameters.

5. Patients with heart disease: This is another growing subset of people who would benefit from light to moderate exercise but are limited in what they can do. The $iSpO_2$ would warn them if they are over doing it, just as we use pulse oximetry to monitor heart patients during stressful situations or during procedures.

6. Stress testing: George W. Bush's coronary artery stenosis (serious enough to require a stent) was first detected on a stress test. He exercises regularly and keeps fit. His only symptom was exercise intolerance. Stress testing will become (or should become) more and more a part of the routine physical exam in the aging baby boomer population. Clinicians should stress test more people, more frequently since there are now less-invasive treatments that can save and/ or improve lives. Just listening to your heart and checking your blood cholesterol is not nearly enough. It would make sense to have a pulse oximeter on everyone undergoing a stress test to make sure they are going hard enough, yet make sure they do not get into trouble.

Performance

There are more sophisticated tests that are more accurate indications of fitness and cardiovascular health. These involve testing at higher intensity and therefore better approximate competitive conditions.

VO_2 max

The best laboratory test of aerobic fitness is VO_2 max, which measures an athlete's oxygen utilization during maximum exercise on a treadmill, stationary bike, or rowing machine. A tightly fitted mask or breathing snorkel connects to a bulky oxygen reading machine as the test subject exercises at full effort. VO_2 max depends on oxygen delivery (atmospheric O_2, air exchange in the lungs, pumping power of the heart, and arterial blood flow to the muscles) and also oxygen demand by the tissues (mitochondria consume nearly all of the oxygen utilized) [14]. VO_2 max (in mL of O_2/ min) is a measure of the athlete's mitochondrial mass. When corrected for the athlete's weight in ml of O_2/(kg × min), VO_2 max is an estimate of the athlete's mitochondrial density. Mitochondria are a major factor in the oxygen delivery portion of VO_2 max (mitochondria power the heart) and in the oxygen demand (how many mitochondria are in the muscles). In novice or less conditioned athletes, VO_2 max tends to be mostly limited by oxygen demand (less muscle mass and fewer mitochondria in untrained people), which can be improved with frequent vigorous training which builds muscle mass and triggers mitochondrial biogenesis (multiplication). In highly conditioned athletes, VO_2 max depends more on oxygen delivery (namely cardiac output) which can also improve with training and is also powered by mitochondria (Figure 10.6).

Sedentary people will have the lowest VO_2 max [15] (low mitochondrial mass), poorly conditioned athletes will have low VO_2 max (the reason why we train), anaerobic body builders may have a low VO_2 max (bulkier fast twitch muscle has a relatively lower mitochondrial density), and marathon runners may only have a moderately high VO_2 max (their events are less about maximal effort and more about endurance and fatigue

Figure 10.6 Greg LeMond had one of the highest VO_2 max's ever recorded. Greg at his
physical best at the 1986 Tour de France, prior to his shotgun injury, lead poisoning, and
mitochondrial myopathy.
Photography: AFP/Getty Images.

resistance). Athletes with the highest VO_2 max are cross country skiers, bicyclists, swimmers, rowers, and medium distance runners. In general, sports which allow repeated maximal aerobic efforts while engaging more muscle groups will result in a higher VO_2 max.

In the laboratory, VO_2 max is accurate and repeatable and can be charted to monitor
an athlete's fitness progress. It is most practical at universities with sports science programs, Olympic training centers, well-financed professional teams, and elite fitness
training facilities. The downsides are that it should only be tested on healthy subjects
because it can cause severe stress in people with diseased hearts and arteriosclerosis.
Even healthy athletes can find it uncomfortable and have difficulty with wearing the
face mask. The apparatus is bulky and impractical for outdoor road use.

Stress testing

Closely related to VO_2 max testing but with more of a medical screening purpose is
exercise stress testing. Heart disease is very prevalent and the number one killer of
people, but can often go undetected and/or have few early symptoms. The stress test
carefully monitors a patient's heart rate, blood pressure, and heart electrical rhythm as
the test subject exercises on a treadmill or stationary bike for 8 to 12 minutes. The test is
curtailed if the patient has difficulty or has signs of cardiac ischemia or EKG abnormalities. Although the resistance may be increased to further stress healthier people,

it does not really measure athletic performance. The stress test is designed to detect cardiovascular disease, namely coronary artery disease, but also cardiac arrhythmias, heart valve problems, cardiomyopathy, and cardiopulmonary disease. Because there are newer and less invasive treatments for coronary artery disease such arterial stenting, the stress test is becoming a more widely used medical screening exam and should be considered before out-of-shape people begin a vigorous exercise regimen. It is also used to clear patients before major surgery, to make sure they will tolerate the stresses of surgery and post-surgical recovery.

Time trialing

You can put your body through the rigors of maximal aerobic effort out it the open air. Solo time trialing (either cycling or running on a fixed length course) eliminates tactics, teamwork, and drafting. All that is left is the "race of truth," as a time trial stage is called in the Tour de France. If you know a local challenging course with no traffic and no major intersections, you can time yourself during the season and monitor your progress. Although you may train at much different distances than this course, it can still be used as a good test of maximal aerobic effort. The downsides are that it can be influenced by technique and equipment (aerodynamics), ambient conditions, and headwinds.

A record time trial

Entering the last day of the 1989 Tour de France, Greg LeMond was 50 seconds behind Laurent Fignon (the French champion). The Tour is traditionally concluded with a relatively uneventful group finish in Paris, but very unusually the 1989 final stage was a time trial. The setting was optimal for a very fast course: a west to east direction with a tailwind, a slight downhill gradient, and a short distance of 24.5 km (15.2 miles). LeMond would also be testing new aerodynamic equipment (specially designed helmet and handle bars) which were available to all competitors (but many of whom considered this equipment too radical, untraditional, or potentially restrictive of breathing). Faced with a task of having to make up too much time in too short of a distance, Greg refused radio contact with these coaches, knowing he would have to go all out the entire way. Pushing a huge gear of 54×12, he finished 58 seconds ahead of Fignon and won the Tour with the closest margin of all time (8 seconds). His average speed of 54.5 km/h (33.9 mph) would set a Tour de France time trial record that would stand for many years.

Time trialing is still a respected discipline in cycling with world championship trophies and Olympic medals awarded. It has become even more popular due to triathlons and Ironman competitions (Figure 10.7).

Lactic acid clearance

Short powerful bursts of speed and acceleration can win sprints and races, but a better measure of fitness and competitiveness may be how quickly an athlete can recover from these intense efforts. When we exert explosive power, we switch from aerobic efficiency to anaerobic inefficiency, resulting in the buildup of metabolites such as lactic acid. Lactic acid is a contributor to fatigue, muscle soreness, and temporarily

Figure 10.7 Greg LeMond in a Tour DuPont time trial stage. LeMond's pioneering innovations in aerodynamic equipment (helmet, aero bars, body positioning, carbon fiber bike, electronic shifting, and clip-less pedals) are still in use today.
Photography: Mike Powell/Getty Images Sport.

decreased performance. The faster we can clear lactic acid from our muscles, the sooner we can rejoin or up the pace [16]. In the circulatory system, arteries deliver oxygen and fuel to muscles, and the veins take away the carbon dioxide and lactic acid. Veins are lower pressure, nonpulsatile, and slower flowing, making them suscep- tible to clots (deep vein thrombosis, i.e., DVT). Exercise does much to prevent DVT by reducing sedentary behavior, maintaining blood vessel health (endothelial linings), and improving muscle tone that helps push the blood back to the heart. A healthy liver is able on convert lactic acid back into glucose, which is just one way the liver assists in athletic performance. Measuring lactic acid in the blood would require an intravascular sensor or blood draw. Perhaps one day there will be a noninvasive means of detecting lactic acid [17]. In the meantime, athletes can be more in tune with their bodies by knowing what degree of effort causes lactic acid fatigue and the pace and time needed to recover.

Body weight

In exercise infomercials and diet print ads, the indication of success is with dramatic before-and-after pictures. Part of human nature and a major aspect of exercise motivation is expecting improvement in body composition. People like to see less fat and more muscle definition. With less fat, people feel and look healthier with more agility and improved self-confidence. Successful weight loss can be encouraging, but lack of results or rebound weight gain can be discouraging. As many have discovered, temporary food deprivation rarely results in permanent weight loss. When we starve ourselves we can lose lean muscle mass and trigger fat storage hormones. There are successful cases where people have eliminated massive consumption of soft drinks, stopped eating deep-fried food, reduced binge eating, or corrected other major flaws in their eating habits. But for the majority of people, exercise must be increased for long lasting weight loss results [18]. Exercise increases mitochondrial fat burning and increases muscle mass, resulting in a lower body weight "set point" and change in body composition. A standard bathroom scale may not be sufficient to measure success in a diet and exercise program. Most of the body mass is water, so weight can vary with hydration status. Furthermore, a standard scale does not tell the difference between muscle and fat. More advanced scales use bioelectrical impedance to estimate body composition and fat percentage because fat is more resistant to electrical current than water or muscle. For devices that measure impedance up one leg and down the other, belly fat may be underestimated. Results also vary depending on hydration status, sweat, and recent exercise or meals. Other devices that more directly measure abdominal fat impedance are used in weight control clinics [19]. Besides the fat just under our skin, we also have fat deep in our bodies in our abdomens. For this type of fat, girth measurement or clothing size may be more telling. Although it is good to have pride in one's appearance sometimes it can go too far and lead to eating disorders or body image disturbance.

Several sports such as boxing and wrestling have weight classifications that encourage drastic weight cutting to "make the weight" which can be unhealthy. Climbing specialists in the Tour de France are careful not to have excess weight since carrying an extra 5 lbs. up a mountain (in either body fat or upper body muscle) could make the difference between winning and losing a stage. The importance of power to weight ratio dictates the sport-specific ideal body shape of committed professional athletes. For fitness athletes, a more balanced distribution of muscle mass and a more realistic body fat percentage is more functional and practical for everyday life.

Wattage

The fastest average speed (crossing the finish line first) wins races. However, during the race real-time speed is influenced by wind resistance, gravity, and tactical pacing. When it comes to physical conditioning and winning, power is what is most important.

Sprinting, fast hill climbing, and breakaway performance require power. This power comes from our cardiovascular system and muscles, and they are in turn supplied by the energy generated by our mitochondria. Electronic strain gauge sensors detect driveline torque on a bicycle resulting in power that is calculated and recorded in watts (a unit of power). Once the purview of only wealthy or well-funded elite athletes, wattage training (i.e., power metering) is now more affordable and more widely available. An athlete's wattage output is an excellent measure of aerobic power, physical fitness, and training effectiveness. A cyclist can expect gradual increases in wattage output with correct training, nutrition, and recovery. In the near future, the technology may be miniaturized and applied to shoes and gloves, making power data available to runners and swimmers too.

Detecting dopers

Performance enhancing drugs are unethical shortcuts to improved power. Anabolic steroids can increase muscle mass, and erythropoietin can increase red blood cells and boost mitochondria to unnatural levels. In the past, cheaters have used science to keep a step ahead of detection. Perhaps now is the time to use science to make sports fair and clean. One idea is to use biometrics by recording VO_2 max and/or wattage of athletes. The power needed for an athlete of a known weight to climb a known height in a known amount of time can be calculated. If athletes show dramatic increases in performance or if race times are incongruous with expected biometrics, further investigation into the possibility of doping may be indicated [20].

Combined electronic gadgetry

New cycling computers, miniaturized sensors, GPS technology, and ingenious smart phone applications can now simultaneously measure and record many types of data. This information can be saved, downloaded, and analyzed.

Maximum speed

If you have a favorite downhill it can be thrilling to attempt a high speed run. A bicycle computer can record your top speed without your needing to take your eyes off the road. Racers can take risks and use aerodynamic positioning to gain on opponents with high speed descents. However, as a single data point, maximum speed does not say much about overall fitness.

Average speed

Some bicycle computers can eliminate minimum speeds (such as below 5 mph) so that very slow speed or stopping at intersections and rest stops is not included. This gives a better representation of training average speed. Average speed increases with improving fitness but also depends on headwinds, drafting, teamwork, and course topography. In a real race where there is no stopping, the shortest average speed means the fastest time and the winner.

Climbing altitude

Some advanced bicycling and running computers can approximate climbing altitude by detecting the decrease in barometric pressure when we ascend. When graphed, the topography of a finished course can be compared with other courses. An excellent way to train is to find challenging hills and climb them at a fast pace. Climbing hills is an excellent way to break through fitness plateaus. Recording altitude helps motivate some athletes to train harder.

Cadence, heart rate, and oximetry

These and other measures can be used individually or combined in graphical analysis to monitor performance improvement and find weaknesses that need more attention. This can result in information overload and obsession with numbers rather than real fitness. Just remember that training and sports should be enjoyable. Use data to motivate and see results, but do not let them dominate your day outside.

Sustained power

In the future, more and more athletes will be able to afford and integrate power metering into their training regimens. Knowing your power output, knowing how long it can be sustained, and how quickly you recover from intense efforts will lead to better understanding about how your body is energized and responds to training.

Self-perception of effort

When athletes spend many hours or seasons training, they become sensitive to the signals their own bodies send them. If you know what to sense and what these changes mean, you can be a very good judge of your performance without requiring electronic gadgetry. Sometimes taking a low tech approach is the best way to get in touch with your body (Figure 10.8).

The single speed bike

The ultimate in low tech bicycle equipment is the single speed bike stripped of all electronic paraphernalia. Your very first bike was no doubt a single speed bike and you probably felt empowered by the increased range and new adventures it provided. Training as an adult on a single speed bike has many physical and psychological advantages. It frees the mind from anticipating what gear you should be in. Instead, your legs adapt to the terrain, connecting you more to the bike and road. You are less distracted by the mechanics of riding and are more aware of your surroundings. You will have to climb steep hills in a higher gear than on your multi-geared bike, but your muscles adapt. Some find that the extra challenge of climbing hills with only one gear helps technique by maintaining momentum, encouraging smooth application of power, and improving out of the saddle pedaling efficiency. Spinning fast on downhill sections can improve cadence and pedal stroke. As a city or commuting vehicle, a single speed bike is easier to maintain and is more durable than a multi-geared race bike. Some of the best cyclists such as Bradley Wiggins and Mark Cavendish began their careers on the track, riding single speed machines.

Figure 10.8 Minimalist single-speed bike (Bianchi Pista).
Photography: Mark Hom.

Advantages of solo training

Greg LeMond broke tradition in many ways. One way was his preference of solo training over team training rides. When you are riding at someone else's pace, it is unlikely
that you are going at your own optimal training pace. To be fair, there were very few
who were at Greg's fitness level at his peak, but even noncompetitive athletes can benefit
from solo training. Many bike clubs have rides with no drop policies. This is to keep
all members intact for safety reasons and to encourage riders of all levels to participate
without fear of becoming lost and abandoned. But riding in large groups makes it harder
for car traffic to get around and pace line riding decreases visibility and reaction time,
risking car related and non-car-related crashes. When you train solo you have the whole
road in front of you and no worries behind you. By sprinting, climbing, recovering at
your own pace, and motivating yourself, you can get the most out of solo training.

Breathing rate and volume

Unlike heart rate, our breathing is easy to observe without monitoring equipment.
When exercising at low intensity you should be able to talk freely or even sing. As
you exercise more moderately you cannot sing and your conversation becomes more
abrupt. At high intensity you cannot speak and are breathing rapidly. It is important

not to pant and rather focus on breathing deeply to make full use of your lung capacity. At very high efforts, your breathing becomes louder as the velocity of the air in your airways increases. Exercise and training will help dilate your airways and improve breathing efficiency. With smoking, bronchitis, and exercise induced asthma, breathing at high intensity can be impaired. During sustained high intensity the metabolism switches to anaerobic metabolism with accumulation of CO_2, lactic acid, and blood acidity. As you ramp down the pace from high intensity, you will find that there is still a lingering urge to breathe hard. The urge to breathe comes mainly from CO_2 build up in the blood stream, but also the need for O_2 to recover and make the switch back to aerobic metabolism.

Foot pressure

Sophisticated power meters electronically detect torque in the driveline of a bike, but you can sense the same forces on your feet. The more pressure you feel on the soles of your feet on the down stroke, the more muscle fibers engage during each muscle contraction, and the more power and torque you are applying to the pedals. This is one way to pace yourself up a challenging hill. Start with low foot pressure in a seated position. This will power you up the base of the hill using efficient aerobic metabolism which is fatigue resistant. Focus on deep breathing and O_2 delivery as you gradually increase foot pressure to increase the power. This transitional phase is where you will discover your lactate threshold (your maximum power level while still being aerobic). As you near the crest, get out of the saddle and apply more body weight into each pedal stroke, engaging more muscle fibers per stroke. This final surge in power will get you to the top faster. During a long race you may at times have to conserve energy with a lower foot pressure (with drafting and teamwork) in order to be prepared for a competitor who may at any time sprint ahead. You cannot maintain high foot pressure all the time as it is fatiguing and uses up stored glycogen. Being in touch with foot pressure helps you control, deliver, and conserve your power. Jacques Anquetil, a five time Tour de France champion, said that his technique was to soft pedal every 5th pedal stroke. Momentum would prevent the bike from slowing down, the momentary ease in muscle tension would allow greater arterial blood flow, and the muscles would have a micro rest. Although this technique is very difficult to emulate, it does have physiologic explanation and showed that this major champion was very much in tune with foot pressure.

Greg LeMond's pedal stroke

In "Greg LeMond's Complete Book of Bicycling" (1988) with Kent Gordis, Greg revealed the secret of his pedaling technique. Whereas most cyclists' power came just from the down stroke using the quadriceps and calf muscles (sometimes supplemented by inefficient pulling up on the upstroke), Greg's unique pedaling technique utilized a pullback motion at the bottom of the stroke. Other cyclists considered the bottom of the stroke a dead zone since pushing down at the bottom of the stroke wasted energy and did not contribute to forward motion. However, Greg's pullback method allowed him to use the powerful hamstring muscles and hip extensor muscles in the back of his thighs (the same muscles used in running). Undetectable to those around him, he just appeared to have a normal (albeit smooth) pedal stroke. This pedaling style and proper bike setup (namely the correct seat height required to execute this style) is another example of Greg's innovation in the science of cycling.

Muscle soreness

When you are training hard, some muscle soreness is to be expected and is not necessarily a bad thing. The soreness comes from lactic acid and other accumulated metabolites, swollen muscle tissue, micro muscle fiber tears, edema, and lymph fluid. As long as the soreness is symmetric, diffuse at the belly of the muscle, gradual, and dull, it usually indicates that you are training hard enough to get fitness benefit and rebuilding stronger muscle. A sharp pain can indicate a muscle tear and pain at the attachment of muscles can indicate tendonitis. The white connective tissue of your tendons and ligaments has less blood flow and takes much longer to heal compared with muscle tissue. A sudden asymmetric muscle cramp can indicate over exertion, dehydration, or electrolyte imbalance. Adequate hydration, electrolyte replacement, timely refueling, and dietary protein can lessen muscle soreness and speed recovery. Post event massage presses excess fluid and metabolites out of the muscles via the veins and lymphatic system towards the center of the body. Frequent or worsening leg soreness with exercise can be a sign of inadequate blood flow due to arterial narrowing or occlusion (peripheral vascular disease). This can be evaluated by a doctor who will check peripheral foot pulses and order other imaging tests if needed. Mild calf tenderness is common in athletes, but calf pain and swelling can be a sign of DVT. The risk factors for DVT include prolonged sitting (long car trips and airplane trips), sedentary lifestyle, certain hormones and birth control pills, and genetic blood clotting disorders.

Time to recovery

Sprinting one time or climbing one hill is usually not that difficult. A better measure of fitness is how fast you can recover from this effort and how many times you can sprint or climb with high intensity. If you are really training with high intensity, you cannot keep it up for too long. Soon you will become very short of breath and your muscles will fatigue. A common mistake on group rides is to pull at the front too hard and for too long, causing exhaustion. Remember that exercise itself does not make you stronger, it is recovery and rebuilding that makes you stronger. With training your heart and muscles will become conditioned and more efficient. You will be able to work harder and recover faster between high efforts. Again, a great way to judge your recovery time is to train solo and go at your own optimal pace and not someone else's pace.

Fatigue (physical and mental)

It would be nice if we could just power along at high speed all the time, but the fact is that our bodies have limitations. We can exercise at low intensity for many hours, but can only exercise a high intensity for relatively brief moments. As you train, your cells grow more energy producing mitochondria, your heart becomes a stronger pump, and your metabolism becomes more efficient. Part of that efficiency is maintaining a higher pace with less fatigue by relying more on oxygen and fat burning. Another is by clearing metabolites with a healthy circulatory system and metabolic organs such as the liver and kidneys. Being aware of your limitations can often tell you more about your fitness than absolute speed or power. Mental fatigue can result from boredom, overtraining, sleep deprivation, and lapses in motivation. It is always a good idea to get adequate rest, exercise with variety, try new courses and venues, and create new challenges that will not only exercise the body but will also stimulate the mind.

Sweating

How much time or mileage does it take to get a good aerobic workout? That largely depends on intensity, but even then there are some minimum exercise requirements in order for you to see the benefits. If you only live a mile from your job, riding a bike to work may get you there in comfort and in dry clothing, but is not likely to be enough exercise for you to see substantial benefit. As many coaches have said, unless you break a sweat you are not working hard enough. When we exercise, some of the energy goes into motion, but as with an automotive engine, much of the energy is wasted as heat. To prevent overheating, we sweat and cool the body with the process of evaporation. As the fastest moving water molecules leave our body through evaporation, they take away the heat. Sweating is yet another process controlled by the autonomic nervous system. We can readily sense when we are sweating and it is indeed a good indication that we are exercising sufficiently. Fortunately, the days of dripping with sweat in a cotton Tee shirt are over and we now have modern synthetic fabrics designed to wick the sweat away, both improving the evaporative cooling effect and increasing comfort.

Listening to your body, being "in the zone"

To conclude this chapter, athletes need to know the advantages and disadvantages of the different methods to gauge fitness. Technology has given the athlete more and more useful information and advances have made this technology affordable and prac-tical. These data can be used to maximize training efficiency, point out weaknesses, and motivate by showing improvement. However, there are times when you might want to eschew the fancy gadgets and listen to your body instead. As you get fit and learn what to listen for, you will get in the zone where your mind and physical body communicate. This more intuitive appreciation of your fitness will make your training fruitful and also more enjoyable.

References

[1] Homer (circa 8th century BC). The Odyssey. Various translations.
[2] Newton I. Philosophiæ Naturalis Principia Mathematica. Cambridge: University of Cambridge Digital Library; 1687.

[3]   Sachdeva S, Davies KJ. Production, detection, and adaptive responses to free radicals in exercise. Free Radical Biology and Medicine 2008;44(2):215–23.

[4]   Sluiter JK, van der Beek AJ, Frings-Dresen MH. Work stress and recovery measured by urinary catecholamines and cortisol excretion in long distance coach drivers. Occupational and Environmental Medicine 1998;55(6):407–13.

[5]   Marinho DA, Barbosa TM, Reis VM, et al. Swimming propulsion forces are enhanced by a small finger spread. Journal of Applied Biomechanics 2010;26(1):87–92.

[6]   Einstein A. Relativity the Special and General Theory: A Clear Explanation that Anyone can Understand. New York: Crown Publishers; 1952.

[7]   Marsden B. Watt's Perfect Engine: Steam and the Age of Invention. New York: Columbia University Press; 2004.

[8]   Tanaka H, Monahan KD, Seals DR. Clinical study: exercise testing, age-predicted maximal heart rate revisited. Journal of the American College of Cardiology 2001;37(1): 153–6.

[9]   Doering LV, Dracup K, Moser DK, et al. Hemodynamic adaptation to orthostatic stress after orthotopic heart transplantation. Heart & Lung: The Journal of Acute and Critical Care 1996;25(5):339–51.

[10] Lakatta EG. Length modulation of muscle performance: Frank-Starling law of the heart. In: Fozzard HA, editor. The Heart and Cardiovascular System. New York: Raven Press Publishers; 1992, pp. 1325–51.

[11] Helvaci MR, Seyhanli M. What a high prevalence of white coat hypertension in society! Internal Medicine: Japanese Society of Internal Medicine 2006;45(10):671–4.

[12] Fagard RH. Exercise is good for your blood pressure: effects of endurance training and resistance training. Clinical and Experimental Pharmacology and Physiology 2006;33(9):853–6.

[13] Vormbrock JK, Grossberg JM. Cardiovascular effects of human–pet dog interactions. Journal of Behavioral Medicine 1988;11(5):509–17.

[14] Wagner PD. A theoretical analysis of factors determining VO_2 MAX at sea level and altitude. Respiration Physiology 1996;106(3):329–43.

[15] Vance DD, Chen GL, Stoutenberg M, et al. Cardiac performance, biomarkers and gene expression studies in previously sedentary men participating in half-marathon training. BMC Sports Science, Medicine and Rehabilitation 2014;6(1):6.

[16] Devlin J, Paton B, Poole L, et al. Blood lactate clearance after maximal exercise depends on active recovery intensity. Journal of Sports Medicine and Physical Fitness 2014;54(3):271–8.

[17] Ren J, Dean SA, Malloy CR. Noninvasive monitoring of lactate dynamics in human forearm muscle after exhaustive exercise by (1) H-magnetic resonance spectroscopy at 7 tesla. Magnetic Resonance in Medicine 2012;. [Epub ahead of print].

[18] Stunkard AJ. The current status of treatment for obesity in adults. Research Publications - Association for Research in Nervous and Mental Disease 1984;62:157–73.

[19] Deurenberg P. Limitations of the bioelectrical impedance method for the assessment of body fat in severe obesity. American Journal of Clinical Nutrition 1996;64(3 Suppl): 449S–52S.

[20] Zorzoli M. The Athlete Biological Passport from the perspective of an anti-doping organization. Clinical Chemistry and Laboratory Medicine 2011;49:1423–5.

BEAST Fitness Training

11

Fitness goals and abilities differ greatly between recreational and professional athletes. However, all athletes have the same biology and thus all can benefit from mitochondrial-based training. You do not have to be a World champion or doctor to benefit from "SciFit" training. You just need to have the desire to improve your fitness. In this chapter, we will offer some tailored training advice for different athletes with different goals (Figure 11.1).

Before we enter into specific training categories, it should be explained what they have in common (varied in proportion and emphasis, depending on the fitness goals):

1. High-intensity interval training (HIIT) in order to trigger mitochondrial biogenesis. With more mitochondria, the athlete will have more energy on tap. With this power the athlete will go faster and endure longer.
2. Frequent exercise to avoid mitochondrial atrophy. With mitochondrial energy, it is use it or lose it. To maintain fitness, we must exercise at least every other day. If we stop exercising for a week or more, we can lose massive amounts of fitness that will take many weeks or months to regain.
3. Base mileage to promote endurance and mitochondrial fat burning. When beginning a new sport or starting a new season, a foundation of low to moderate levels of exercise is especially important to prime the body in preparation for the more intense training to follow.
4. Strength building with resistance exercises to build muscle mass. This will raise metabolism, increase power, change body composition (more muscle and less fat), improve outward appearance, and lessen or prevent age-related muscle atrophy.
5. Avoidance of overuse, overtraining, and injury. Too much exercise of the wrong type can be counterproductive. Any serious or nagging injury can be a major impediment in attaining fitness.
6. A balanced diet including: fuel for cellular energy, hydration for optimal performance, and nutrients for the recovery and rebuilding of our cells and muscles.
7. The motivation to stay with a sport or exercise program for years (or a lifetime). Exercise should always be a pleasure and should never become a chore.
8. The goal of ever improving fitness. Although an Olympic athlete knows what it is like to reach the highest levels in a sport, most of us fall well short of reaching our true athletic potential. SciFit training advice will try to raise the athletes' bar, so they can expect more and get more out of their training.

Recreational fitness training

The recreational athlete faces the challenges of fitting exercise time into a busy schedule, often having a lower baseline fitness level, and often having fewer resources or support. The recreational athlete may have to deal with medical conditions such as

The Science of Fitness: Power, Performance, and Endurance. http://dx.doi.org/10.1016/B978-0-12-801023-5.00011-9

Figure 11.1 Training principles apply to competitive and noncompetitive athletes. Amateur race along the James River in Richmond, Virginia.
Photography: Alan Cooper, President of the Richmond Area Bicycle Association (www.RABA.org).

being overweight, high blood pressure, or high cholesterol. SciFit training is helpful to the recreational athlete since it offers condensed high intensity workouts, which are more beneficial and less time-consuming than all-day, long-distance training. Frequent exercise with gym membership, home equipment, or bicycle commuting prevents fitness relapse common in the weekend warrior. Progressively harder workouts will boost mitochondria and overall health.

If you ask recreational athletes why they exercise the answer is often, "For more energy." That is clearly what we are after with this book. More mitochondrial energy for vitality and performance. They may also say, "For health." Although you will have no trouble finding a physician to prescribe pills to treat hypertension and high cholesterol, or a surgeon to staple your stomach if you are obese, the first line treatment is diet and exercise. Although the phrase *diet & exercise* has been overused to the point of being cliché, we hope this book offers useful insight on how to create a healthy lifestyle. If a measure of health is no longer requiring medication to treat preventable diseases, then yes, becoming fit could make you drug free and much healthier. They may also say, "For fun!" When you are engaged in a physical activity that you enjoy, mental stress is relieved, stress hormones abate, your mind is sharper and more focused, you sleep better and awaken refreshed, and you have more energy to enjoy time with your family and friends.

"You never told me it was fun"
Interventional radiology is a gratifying profession because we can make a difference. By using state of the art imaging we can see deep inside of patients to perform precision biopsies or effect direct therapies, often seeing improvement occur right before our eyes. Ordinarily medical practice is prescribing medication and waiting for change that may or may not happen. That is why interventional radiology is worthwhile. However, it is also a stressful job where lives depend on your decisions and where medical errors and bad outcomes are not acceptable. One way I handled the stress was to ride my bike to and from the hospital with a 12 mile commute each way. I would arrive at work or at home mentally and physically charged, but also unstressed. A coworker of mine was feeling the stress. He had gained weight, was on blood pressure medication, had trouble sleeping, was headed for metabolic syndrome, and barely had enough energy to make it through some of our longer days. At my urging he decided to try bicycling and he loved it. In a few months he built up his fitness and endurance and was soon bike commuting daily. We started a competition with a tally board that counted every day we rode a bike to work. There were some days when it would have been easier to just jump in the car, but I rode just to have another hash mark on the tally board. In a good year we would each bike commute 100 times. In a few months my coworker lost 40 lbs., lowered his blood pressure enough to no longer require medication, lowered his mental stress, and regained his all day energy. His wife told him that he stopped snoring (which can be a sign of sleep apnea); he now wakes up refreshed instead of tired. Despite all of these health benefits he said, "You never told me it was fun." The real reason he bicycled so much was because he got pleasure from it. When we engage in strenuous aerobic exercise, endorphins (natural opiates) are released and bind to pleasure receptors in the brain. Endorphins induce calmness, alleviate pain, and can give a feeling of pleasure and euphoria. Often called the "runner's high", this state of pleasure can be achieved with other strenuous aerobic activities too. My coworker was also spending more time outside riding with his grandchildren. Social interaction and outdoor sun exposure can elevate mood and brain activity. Another reason he was happy was that he felt better about himself, both in physical appearance and in retaking control over his own health.

– Mark Hom

Weight control

Overweight people often have a sedentary lifestyle coupled with poor dietary choices that can result in muscle atrophy and excessive body fat. To combat metabolic syndrome, obesity, and type 2 diabetes, a fitness plan designed to rebuild muscle mass and burn fat is essential. Getting rid of excess fat can extend your lifespan, dramatically improve your quality of life, and motivate you to continue exercising. We are learning more and more about body fat: the limitations of body mass index, the health implications of where our body fat is located (deep truncal obesity is especially dangerous), the importance of brown fat in adults (proven with positron emission tomography scans), and that mitochondria are at the center of the fat burning process. If you are to believe the infomercials, fat burning comes in a bottle (a miracle pill).

In reality, fat burning (beta oxidation) comes from inside of your cells, from your mitochondria.

Training to increase lean muscle mass also increases your mitochondrial mass. Because mitochondria perform nearly all of the fat burning in your body, there is a double benefit when exercising to lose weight. You not only burn fat while exercising, proper exercise creates more mitochondria which will raise your metabolism. A slow walk around the lake is not enough to boost your mitochondria. Vigorous muscle-building exercise is required. Attention to a healthy diet is also critically important. Some vitamins, minerals, and nutrients can enhance mitochondria and, therefore, assist the fat burning process. However, there is no such thing as fitness in a bottle, and no substitute for a sensible diet and vigorous exercise when it comes to getting into, and staying in, great shape (Figure 11.2).

Figure 11.2 Greg LeMond preferred solo training where he could push himself at his own pace. Long distance team rides were a waste of his training time. Greg in the stars and stripes jersey of the USA world championship team.
Photography: David Madison/Getty Images Sport.

Seasonal training

Most athletes do not live in areas that permit outdoor exercise 365 days of the year. In addition, certain times of the year, winter in particular, are associated with a "seasonal loss of fitness." Ideally, we would stay on top of our game throughout the year, but we are not perfect and seasonal loss of fitness is the reality most athletes must face, even among the most avid. Not only do we build mitochondria with exercise, we also lose them with inactivity; however, we can mitigate this loss with off-season training. The best way to deal with seasonal loss of fitness is to view the year as a cycle. Try to be active during the off season either by using a home aerobic machine, weight training, or adopting a new winter sport. Even extra stair climbing at work will help you preserve some of your fitness and prevent muscle atrophy. When the off-season finally ends, you will be in much better shape and have many more mitochondria than those who stopped exercising all together. In the early season, fitness should be seen as a step-wise building process (as you build up more mitochondria and strengthen your muscles). In the prime season, you should continue to improve and not plateau, by challenging your body with more intense workouts. As the season winds down, you should stay active to maintain the fitness you have gained.

Clavicle and Five Ribs

Riding with my bicycle club on a Saturday morning ride in June, I had just taken my turn at the front of a long pace line and was filtering back at close to 30 mph, when our group was passed by two speeding cars. I thought at least those reckless drivers are past us and won't be bothering us anymore. However, what they were attempting to do was save 20 seconds of time by beating us to their driveway. Because they had to speed to pass us, they overshot their driveway and came to a stop, blocking the road with one of the cars as they "J-hooked" our cycling group. This caused a multi-bike crash with our then club president suffering a concussion and me flying in the air and landing hard on the asphalt, shattering my left clavicle and five ribs. When I got home from the Emergency Room, the pain of just breathing or rolling over in bed was excruciating. I knew it would take 6 weeks to heal, but also knew that if I were to lie in bed for 6 weeks my muscles would atrophy, my mitochondrial energy would diminish, and my general health would decay. I knew I had to get moving. At 2 weeks I was still in pain but forced myself to go to work. It was too painful to ride a bike or run, but at 3 weeks I started climbing the stairs at the hospital where I worked, 12 stories from basement to roof level. I skipped every other step and walked up as fast as I could. Soon I was doing the double, 24 stories. At 6 weeks I was healed enough to get back on the bike. I spent a week riding solo and bike commuting to overcome my fear of reinjuring myself. On my first Saturday morning club ride since the accident, my wife (who never had to stop training) thought she finally had a chance to beat me to the coffee shop. Not only did I get there first, but I broke away and beat the entire group to the coffee shop. I had my coffee in hand as my wife and the group pulled in, much to her chagrin.

Mark Hom

Hill climbing training

Whether it is the Tour de France or a local club ride, there is one challenge that separates cyclists the most: a long steep hill climb. When people first try cycling, hills are associated with exhaustion and discomfort and are often avoided. Once you get some miles in your legs, you will find that a long steep hill is your best training partner. Fighting the unrelenting pull of gravity offers the perfect resistance to put your muscles into the energy sapped state that boosts mitochondria. Fast paced climbing and hill repeats are the best methods to trigger mitochondrial biogenesis. At the base of a hill, you can start with a high cadence and low pedal pressure to stay in a comfortable aerobic zone. This method of hill climbing also avoids pumping up the muscles which can restrict arterial blood flow. However, if the pace increases or the climb steepens, you must increase pedal pressure. This incorporates more muscle fibers per pedal stroke and this "horsepower" can be accurately measured with an on-bike wattage meter (either hub, crank, or pedal based). Your heart rate and respiratory rate will start to soar. In lieu of monitoring devices, you can rely on perceived pedal pressure and your breathing as indications of the effort you are expending. As you surge on the hardest part of the climb, you are out of the saddle applying maximum pedal force and breathing hard and fast (Figure 11.3). Although you are now in an anaerobic zone and nearing exhaustion, you are also triggering your mitochondria to multiply. In a few days, you will find yourself going up that same hill at a faster pace.

Figure 11.3 Greg LeMond (second in the row) climbing in a mountain stage in the 1989 Tour de France.
Photography: AFP/Getty Images.

Hill repeats are an excellent and practical way to boost your mitochondria. Just find a challenging local hill on a loop with little traffic. After a warm up lap, try attacking the hill at an ever increasing pace each lap. For a real challenge, try riding up the hill using only your big front chain ring (higher gear range). Use the downhill section to recover for the next lap up the hill. This method not only triggers mitochondrial biogenesis many times (on each uphill section), it also encourages rapid clearance of lactic acid during the downhill sections. For variety, try different terrain contours. Good examples include a hill which gradually steepens (to test your pacing) or a hill with multiple steps in gradient (which can be attacked with brief but intense bursts). If you must train after work, there may be too much traffic on the major roads during the rush hour, but you may be able to find a less trafficked hill circuit near your house that will allow for an intense hill repeat workout. If you train on hills, you will see benefits when riding on flat roads too. If you race or ride with a club, you will find that you can breakaway from the pack at will, or when you see riders ahead you will find that you can chase them down with less effort. Remember that gravity and acceleration are the same. Training against gravity will make it easier for you to accelerate and breakaway, on any terrain.

Power training

Performing with power requires a combination of aerobic and anaerobic fitness that can be accentuated with special methods. An example is a single speed bike which encourages the muscles contract more forcefully. Instead of down shifting to an easy gear when climbing a hill, a single speed bike forces you to ride up that hill with high intensity. At the top of the hill you will be exhausted and breathing hard, but satisfied knowing you are triggering mitochondrial biogenesis. The next time you ride your multi-speed bike, you will be stronger and faster. If you feel that you have reached a fitness plateau on your regular road bike, a single speed bike can help you break through that ceiling. Sprinting is a great way to boost muscle strength and mitochondrial power. An example is Mark Cavendish who began his career as a track cyclist and now has tallied an extraordinary number of Tour de France stage wins (Figure 11.4). Sir Bradley Wiggins (the 2012 Tour champion) was also a track cycling star who strengthened his legs on a single speed bike. Measuring oxygen consumption on the bike is not practical and heart rate monitoring has its limitations. However, power metering can give the athlete real time readout of his output in watts. This technology is miniaturizing and becoming more affordable to everyday athletes, making wattage metering an increasingly important part of training and fitness coaching.

Time trialing is a discipline that builds sustained high levels of power. Although commonly identified by the wind-cheating aerodynamic equipment (pioneered by Greg LeMond) time trialing is really about human power output. To succeed at time trialing you need the same qualities that are measured with VO_2 max: a high cardiac output and high oxygen demand from your muscles.

Figure 11.4 Mark Cavendish (in the middle), 2011 World Road Cycling Champion. Photography: Mogens Engelund.

Endurance training

A common misconception is that long steady distance (LSD) is the best way to train for endurance events such as a marathon, century ride, or triathlon. Long-distance competitors often train for their events by gradually lengthening their training distance, but maintain the same slow to moderate pace. Compared to training at a jogging pace, HIIT will do more to trigger mitochondrial biogenesis. Endurance athletes can benefit by integrating periods of intensive effort, that is, interval training into their workouts. This will increase the number of their mitochondria, which in turn will supply them with extra energy they will need when participating in an event. Even for the endurance athlete, quality (high intensity) training can boost mitochondria more than quantity (high mileage) training. Many long-distance runners are reluctant to give up on the idea of LSD training because of lore and tradition. Another reason is because the physical act of running causes foot and joint impacts that require recovery on "easy days." They should be aware that running with sore tendons and joints is a setup for overuse injury. Either nonimpact training or a day of complete rest would be better than running in pain. LSD training has one major benefit in that it encourages fat burning over glycogen utilization. Fat is very energy dense and a very efficient fuel source for your mitochondria. Glycogen, on the other hand, is a more limited fuel source that is best reserved for shorter bursts of speed.

Resistance training

Resistance training has multiple benefits including: adding muscle mass, stimulating bone growth, improving basal metabolism, improving physical appearance and self-esteem, improving function for everyday activities, maintaining fitness during the off season, fighting against age-related muscle loss, and increasing independence as we get older. Although you should exercise at least every other day, inclement weather does not always permit outdoor training. There is plenty you can do indoors to maintain or build your strength. When it comes to resistance training most people think about using machines or free weights at the local gym; however, there are some resistance exercises you can perform conveniently at home using minimal or inexpensive equipment.

Super-set count-down high-volume training uses your own body weight to build strength and to increase fatigue resistance. Super-sets are exercises that use opposing muscle groups which allow you to work the different muscles immediate after each other which gives some cardiovascular benefit and also keeps the opposite muscles groups well-balanced. Counting down the number of repetitions (reps) per set accounts for normal muscle fatigue. Fewer and fewer reps are performed as the workout progresses. High volume means that the total number of reps is high, to improve both fast twitch and slow twitch muscles. It literally means that your muscles are doing more "work." One example of this method is a workout combining pull-ups and push-ups. Pull-ups (and chin-ups) strengthen the biceps, latissimus dorsi, and back muscles. Push-ups strengthen the triceps, pectoralis, and deltoid muscles. If you can do 12 pull-ups maximum in one set, start at a lower number such as 10 pull-ups, followed by 20 push-ups with no rest in between. Then count down and do eight pull-ups and 16 push-ups. Then count down and do six pull-ups and 12 push-ups, and so on.

Your reps counts will look like this:
10 pull-ups	20 push-ups
8 pull-ups	16 push-ups
6 pull-ups	12 push-ups
4 pull-ups	8 push-ups
2 pull-ups	4 push-ups
For a total of:	
30 pull-ups	60 push-ups

By counting down you can still perform the exercises as your muscles naturally fatigue, yet you will achieve a high total number of reps, much higher than your one set maximum and a lot more work. At the end of this workout, your heart will be beating fast and your upper body will be pumped. It is a deceivingly good workout. A few tips: always perform the movements with smooth control and good form to prevent jerky motions that can cause injury. Stop immediately if you feel any tendon or joint pain as the high volume could aggravate tendonitis. Do not try to lift your chin above the bar at full strain as this could injure your neck. Lifting yourself so your eyes are even with the bar is sufficient. Once you have done this workout every other day for 2 weeks, increase the starting sets to 12 pull-ups and 24 push-ups for 42 and 84 total reps respectively. This ramping of the challenge to your muscles can increase indefinitely and your muscles will respond by growing stronger and by developing increased resistance to muscle fatigue. You may not become huge and muscle-bound, but it will give you practical strength that will allow you to shovel the entire driveway or rake the entire yard without exhaustion. If you cannot do enough pull-ups to begin this method, you can use an inclined pull-up machine such as the *Total Gym* that uses just a fraction of your body weight with the same beneficial range of motion.

A simple toning exercise for your lower extremities can be done nearly anywhere, such as a hotel room, dormitory, or office. Stand with the ball of one foot on a short platform (a thick book will do) and grab the back of a chair for balance. Allow your heel to drop. This is the starting point of a single heel press. But also bend your knee about 20°. Then press your body up vertically, lifting your heel and straightening your knee simultaneously. This engages the calf muscles (gastrocnemius and soleus), thigh muscles (quadriceps), and hip extensors. Repeat until you feel mild muscle burn, then do 10 more, then switch to the other leg, then count down and do 10 fewer reps, and repeat. If you can build up your strength and endurance to do 80 reps on each foot, the next set is 70 on each foot, then 60 on each foot, and so on. The countdown decrementing reps allow for a high total volume of exercise but take into account normal muscle fatigue. Although the move seems simple, it engages major muscle groups in your legs and also requires core and hip muscles to prevent the offside hip from sagging. If you do this move on a leg press machine with your back pressed against the seat, you will not get the same core benefit. This is another way to take advantage of gravity and normal muscle fatigue to build strength and fatigue resistance. Soon you could be starting at 100 reps. Take care not to go too far and aggravate foot problems or tendonitis. In general, this method is less injurious than powerlifting or running because it encourages high rep counts at low strain with no landing impact. It strengths both fast and slow twitch muscles, increases resistance to fatigue, improves calf muscle tone to prevent deep vein thrombosis, and causes the muscle exhaustion that triggers mitochondrial biogenesis.

Kettlebell training is an ancient strength building system that is currently undergoing a resurgence in popularity. The kettlebell is a round weight with a handle carved or molded on top. At the Archaeological Museum of Olympia in Athens Greece, there is a huge 143 kg stone kettlebell (a boulder with a handle and an inscription that claims that it could be lifted overhead with one hand). Cast-iron and steel kettlebells made their way to Russia were they were used as farm implements (called "girya") to weigh harvested crops, when it was noticed that the farmers who handled them became physically stronger. About 300 years ago, Russians created kettlebell strength demonstrations, challenges, and competitions. With typical gym weight training, the focus is on muscle isolation, symmetric movements, and restricting momentum (no swinging of the weights). Kettlebell training is quite the opposite. With most kettlebell maneuvers, the weight is lifted asymmetrically with one hand. This engages the stabilizing muscles (the large deep core muscles that connect your lower limbs to your pelvis, your pelvis to your back, and your back to your upper body). With kettlebell training, it is less about isolating one group of arm muscles and more about using your entire body to lift the weight. Many kettlebell movements employ swinging and momentum to lift the kettlebell, which is strongly discouraged as "cheating" with free weights. With kettlebell training, swinging the weight is encouraged because this dynamic motion improves your timing, balance, and coordination. There is a practical benefit since you will become more adept at using one arm to lift your child or a heavy suitcase (most daily chores require asymmetric strength). Because this method of training requires learning new skills, it is essential that you receive training from a certified kettlebell instructor to make sure you are practicing proper technique in order to get the most benefit and to avoid injury.

Age-specific training

If there is ever a time in your life when you will get the most results from exercise, it is in your youth. Around the time of adolescence, hormones such as growth hormone and testosterone are surging. If you exercise at this age, these hormones encourage musculo-skeletal development and mitochondrial muscle distribution that will yield benefits well into adult life. Childhood exercise should be encouraged since active children are more likely to become active and healthier adults. Precautions include prevention of injury from accidents by the use of protective equipment such as bicycle helmets, safety education, avoidance of risky extreme sports (skate boarding down staircase handrails for example), and avoidance of head injury (when playing football and soccer). A child's tendons (muscle attachments) are still developing which is one reason why junior racing bicycle gearing is often restricted. Despite these risks, the dangers of children not exercising are even greater. For example, type 2 diabetes was once called adult onset diabetes because it was a complication of obesity in adults. With children being less active in the modern age, there are now epidemics of obesity and type 2 diabetes in children. The obesity issue is complex as there is ironically a strong association with poverty and obesity. This points to a nutritional problem as well, as convenient and inexpensive foods are often high in calories but poor in nutritional value. Epidemiologists predict that today's children may be the first generation to be less healthy than their parents. Only proper exercise and a correct diet can reverse this negative trend.

Young adults also reach a crucial time in their physical training. As people reach their 30s and 40s, they have many more time commitments and responsibilities such as work, marriage, children, and a mortgage. Often the first thing to be sacrificed is regular exercise. It is about this time in life that many people stop exercising, sometimes forever. They feel healthy and just do not have the time. However, inactivity at this age results in the insidious onset of muscle atrophy, mitochondrial loss, and increased body fat. It is very important to continue exercising at this age, or restart ones exercise routine. Many decide too late after they have stopped exercising for such a long time that they are now too unhealthy or overweight to feel good about exercising again. By exercising as an adult you will not only serve your family by preserving your health, you will also have more energy to fulfill your other responsibilities. Fortunately, many of the SciFit exercise routines are intense in effort but relatively brief in time. For example, a local 45 minute hill repeat ride will result in more mitochondrial boosting than an all day low-tempo ride out in the countryside.

Mature and older athletes must deal with maintaining muscle mass and weight control when hormones are naturally waning. Exercise is the best way to stay healthy, strong, and independent as we age. An older athlete may no longer be able to exercise with the same power and intensity, but moderate exercise will still yield tremendous benefits. The current thinking about the mechanism of aging now is centered on mitochondria, where harmful free radicals are formed and must be neutralized, antioxidants perform their protective duties, and energy is produced throughout ones life. Although some claim to "treat" aging with hormone supplementation (human growth hormone and testosterone), we do not recommend that because a second adolescence at this age is unnatural and these substances are banned by all ethical sports organizations.

Competitive and elite training

Elite athletes have many more mitochondria than average athletes. One observation in any athletic club is that the fast athletes get faster as the season progresses. That is because these athletes have the mental drive to push themselves (thus boosting their mitochondrial energy) to ever-higher levels of fitness. The same mitochondrial principles that work for recreational and club athletes also apply to elite competitors, only taken to a higher level. The competitive athlete must focus on: interval training to boost mitochondria, adequate sports nutrition and hydration, avoidance of overtraining injuries, awareness of the trappings of exercise addiction, and devoting adequate time for sleep and recovery. Although exercise forces your muscles and mitochondria to respond (in good ways), they only rebuild and become stronger when you are at rest. Too much of the wrong exercise, overuse injuries (such as tendonitis), sleep deprivation, and compulsive exercise behavior are counter-productive in attaining optimal fitness. How do you know if you are getting enough sleep? One sign is if you wake up feeling well-rested, before your alarm clock goes off. If not, you need to get to bed earlier. The brain rebuilds its limited glycogen supply as you sleep and needs this down time to reenergize.

Event peaking

Well-planned, long-range, and step-wise improvement of fitness will win the big race. With the correct training and by taking good care of your mitochondria, you will be well prepared for the big day. Tapering (i.e., reducing exercise intensity and volume in the days or weeks prior to a big event) must be balanced between allowing for recovery and not descending into atrophy. There is a great amount of lore about tapering, especially amongst competitive runners, who must recover from micro injuries due to the millions of impacts their joints, tendons, and ligaments receive during training. When performance falters, a running coach will generally prescribe rest. We believe that a better plan would be nonimpact exercise, to mitigate the risk of mitochondrial atrophy.

During the actual event, the athlete's energy must be carefully expended and timed for a good showing at the finish. If you followed the advice in this book, you will have trained with the intensity needed to optimize your mitochondrial power. Many new competitors take this intensity to the race by working too hard at the start, using all of their stored glycogen in the middle of the race, and having nothing left at the finish. To win, the competitor must be savvier. It is better for your teammates to work with you, or better yet have your competition work for you. Competitive athletes must be aware of their biological strengths and limitations, saving their energy for the most critical parts of the race.

The Finish Line...and a Starting Line

Figure 12.1 Greg LeMond winning his third and final Tour de France in 1990.
Photography: Tony Duffy/Getty Images Sport.

Empowered with the knowledge of how your body works, how your cells function, and how mitochondria make your energy, you will be able to optimize athletic power and improve overall health. This book charts the way to a complete, science-based training system, one that will lead to a fuller and more energetic life. The goal of this book was to make physical training more scientific and less dogmatic by explaining why exercise works. Because mitochondria are so pervasive in our bodies and cells and because they supply our energy, understanding them can help us become more physically fit and avoid preventable diseases.

Congratulations! Since you are reading this now, you have reached the end of our book, the finish line so to speak. We hope you found it to be a better explanation of fitness than what has previously been offered. It may not have been easy reading. We packed it with as much useful information as we could and cut out the "fat" to make it lean and concise. Because of that you may have had to skim over some of the more technical and scientific chapters, depending on how much you remember

The Science of Fitness: Power, Performance, and Endurance. http://dx.doi.org/10.1016/B978-0-12-801023-5.00012-0

from high school biology. That is perfectly alright and those chapters will be waiting for you. You cannot be expected to get it all the first time through. If you did read the entire book cover to cover, we hope that you found our explanation of fitness to be deeper than what you would get from an ordinary fitness book. That was our intent, to answer the questions many athletes have on their minds, instead of just telling them what to do.

The concept for this book came about when I (Mark Hom) was trying to get back into shape. I had always prided myself on being in good shape, from riding my bike and strength training. I remained at my lean high school and college weight long after I graduated. When I became a doctor, I fulfilled a dream I had of owning a motorcycle. Because I waited so long before buying one, I avoided learning to ride during the dangerous years of youth and have never had an accident. I found a local group of motorcyclists and took long group rides to different scenic areas of my state, logging many, many miles on that first motorcycle. Motorcyclists can be the most interesting and unique people you will ever meet. One evening while visiting my parents, I weighed myself on their bathroom scale. I told my mother that her bathroom scale must be broken because my weight which had always been 165 lbs was now 172 lbs. But the scale was completely accurate. Body fat has a way of sneaking up on you because much of our fat is deep inside our abdominal cavities (belly fat). The problem was that my infatuation with riding a motorcycle had replaced my bicycle riding. When one of my motorcycling buddies came by, he saw my bicycle collection hanging on the wall. He said they were nice, but asked me why they were gathering dust. He did not believe my explanation that the last road I was on must have been really, really dusty. I took it upon myself to get back on my bicycle and nip this weight thing in the bud. However, I wrongly assumed that my fitness was still there. I am fortunate to live on a scenic road with tree-canopied rolling hills that look great from the seat of a car or the saddle of a motorcycle. But on that first day back on the bicycle, I was not enjoying the scenery; I was agonizing over the hills. I only made it about 2 miles out before I had to dismount and lie down under a shade tree.

In the 2 years since replacing bicycling with motorcycling, I had gained weight and lost all of my fitness. But by continuing to train at longer distances and faster paces, I eventually rebuilt my fitness and got back to my college weight. I spoke to a female athlete who started exercising again in middle age for the cardiovascular benefits and in doing so secondarily lost a significant amount of weight in the process. When she found out I was writing a fitness book, she said that one question she had was, "Where did the fat go?" The answer to her question and the reason why I lost my fitness with inactivity and rebuilt my fitness with exercise can be answered with one word: mitochondria. When we exercise, our mitochondria multiply and burn our fat.

By focusing on mitochondria, I found a way to explain exercise more scientifically with a grand unifying theme. The other fitness books always stopped short. They offered (often arbitrary) advice, but never explained *why* their advice might work. There was certainly no shortage of fitness books, but I thought there needed to be a new one. The reason why fitness books are so popular is because, like me, people have unanswered questions. I remembered how my number one sports hero, Greg LeMond, became the first American cyclist to win the Tour de France in 1986 and how his 1988 book revealed his technical innovations in bike setup and pedaling technique as well as his systematic approach to training. He could out climb, out sprint, and out last everyone. And no one could touch him in the time trial, the "race of truth." Here was a pure manifestation of mitochondrial power, combined with pioneering innovation. When he shocked many by retiring prematurely because of mitochondrial myopathy from lead-poisoning, I thought that his entire career (both the highs and the lows) and fitness in general could be explained by mitochondria (Figure 12.1).

On a lark, I emailed my first few chapters to Greg LeMond. When he responded in total agreement with a 4000 word email, I knew I was on the right track and that I might get a publisher interested in the concept. But it was not quite that easy. If you tell a literary agent that you are writing a fitness book they are not impressed. There are already too many fitness books out there. Not only that, but when you mention a chemical such as adenosine triphosphate, their eyes glaze over. Mainstream fitness publishers expect books to be dumbed down with the usual rehash of traditional lore and unfounded advice. But I think those publishers underestimate the curiosity and intellect of the modern athlete. Today's athletes are eager to learn more and delve deeper into the reasons why physical training works. Looking outside of mainstream fitness publishing, we found acceptance in academic publishing. We were very fortunate to find an editor at Elsevier (the largest scientific book publisher in the world) who thought that a more scientific explanation of fitness was *exactly* what was needed. During the peer review process, we received even more encouragement that we were on the right course. The reviewers completely agreed that our focus on mitochondria was the right one, saying that there was a great need to share this information with the public. They felt that there is currently a gap between biology and fitness that this book could fill. Mitochondrial science is becoming a new branch of medicine with an explosion of mitochondrial research that not only explains physical fitness, but also explains the underlying causes of obesity, metabolic syndrome, cardiovascular health, degenerative disease, and the latest theory of aging.

With this feedback, we expanded the health aspects of the book concept. Mitochondria are pervasive in all of our cells, organs, and tissues. Mitochondria not only explain the biology of fitness, but also explain health and why lifestyle (diet and exercise) can prevent the modern diseases of inactivity and self-neglect. In medical school, I learned how the body works in my first year classes and how various diseases made things go wrong in my second year classes. During the third year, we ventured out into the hospital on our clinical rotations. By luck of the draw my first experience with patients was in the Neurosurgical Intensive Care Unit, which was like learning to swim by diving into the deep end. The chief neurosurgeon was a brilliant teacher and would correlate the surgical and radiologic findings with the neurologic signs and symptoms of our patients. Back then the progress notes were hand written. The most difficult patient in the unit had reams and reams of medical records describing his many problems and treatments. I asked the resident above me, about the cause of this patient's original admission into the hospital. His cynical answer was, "Train wreck." Being naïve, I believed him for a few days. But looking back on what is now a 25 year medical career as an interventional radiologist, the current state of modern medicine is designed to treat the "train wrecks." Certainly we have the task of putting people back together after trauma or serious illness, but what if we could prevent the train wreck in the first place? What if we spent more effort in maintaining the locomotive to prevent catastrophic failure? What if we could keep the tracks in better alignment to prevent derailment? What if we could steer a better course to prevent a devastating head-on collision? As a physician, the best advice I can give is for people to stay healthy and avoid a stay in the intensive care unit. There is plenty we can do with our lifestyle choices to avoid serious illness. The single best thing you can do is to stay active your entire life.

Another hope with this book is to increase awareness about mitochondrial disease. Mitochondrial science is a burgeoning area of new research that is unlocking the secrets of life and cell energy. Mitochondria are at the center of the diseases of modern society and inactivity such as: obesity, metabolic syndrome, and type 2 diabetes. Mitochondria are most concentrated in the most important organs in our bodies: the heart and the brain. Mitochondria can be impaired by toxins and commonly prescribed medications. Mitochondria have a significant role in neurodegenerative diseases such as Alzheimer's disease and Parkinson's disease, and are central in the aging process. Within the limits of a marketable fitness book, we could only touch upon these subjects. One advantage with working with Elsevier is that they had the vision to foresee *two* books stemming from our central theme. "The Science of Fitness: Power, Performance, and Endurance" was written with the athlete and general public in mind, to give an overview of the mitochondrial role in fitness and health and to offer practical training and dietary advice. We are already working on the second book, "Mitochondrial Fitness: The Science of Athletic Energy." This will be a scientific reference book which will delve into the science and research in much greater detail to be a resource for physicians, scientists, and students of biology and physiology, and to further encourage the work being done by mitochondrial researchers. Great strides are already being made by organizations such as the United Mitochondrial Disease Foundation (www.UMDF.org) whose mission is:

"To promote research and education for the diagnosis, treatment, and cure of mitochondrial disorders and to provide support to affected individuals and families."

However, public awareness about mitochondria is still lagging. Mitochondrial diseases remain a challenge to diagnose and treat. Although there is a long way to go before we have complete understanding about the entire role of mitochondria in health and disease, our two books will hopefully advance the cause by increasing knowledge and encouraging further research. Sometimes the finish line is also the starting line.

Dr. Mark Hom, M.D.

Index